CLOUD
SOURCING
THE CORPORATION

THE SINGLE MOST IMPORTANT TECHNOLOGY LEAP
SINCE THE ADVENT OF COMPUTING

BEN
TROWBRIDGE

FOUNDER AND CEO OF ALSBRIDGE

$\mathcal{A}$LSBRIDGE

ISBN: 978-0-615-46799-3

CONTENTS

ACKNOWLEDGEMENTS

I WOULD LIKE TO PERSONALLY thank the many Alsbridge clients who continue to help shape our thoughts on the future of cloud computing. Your insights, belief in Alsbridge and engagement of us as your sourcing and benchmarking advisor is both humbling and exciting.

The foundation of this knowledge and of this book is the Alsbridge team. I want to thank each of them for their tireless efforts to define the possibilities of cloud sourcing and for providing such fair and unbiased advice to our clients – and to me as CEO.

Special thanks to Sean Halverson who worked so hard to develop our understanding of the cloud market, David England for energizing and improving the way to communicate cloud sourcing, Paul Hardy for his patient and insightful shepherding of our team, Pat Garrett for his ability to see far into the future, Barbara DeGuise for the depth of her thoughts, and David Mitchell, Mike McGarry, Mike Thompson and Paul Cervelloni for their contribution to our overall knowledge base along with many others who worked to refine our cloud offering into an industrial strength methodology.

PROLOGUE

THE CLOUD AND CLOUD SOURCING provide one of the most important opportunities for business since the advent of computing. In the development of this book we have focused the best of Alsbridge sourcing experience and provider evaluation techniques to provide an outline of how to effectively engage the right cloud sourcing provider.

As you evaluate the most effective method of engaging cloud services, you will quickly discover the wide range of business applications and infrastructure on demand that will create opportunities to solve business problems where they previously may have gone unaddressed.

Expect a rapid pace of change. The cloud sourcing provider capabilities represented at the date of publishing this work in chapter 8 will rapidly change over the next few months. The increasing velocity of cloud adoption and the already-occurring acquisition of smaller cloud providers will continue to evolve the cloud sourcing services landscape.

The price points of cloud providers promise significantly lower prices for certain services that historically would have been provided by an outsourcing provider. The lower price points and rapid growth of the cloud providers as compared to the low to neutral growth of the traditional outsourcing market is a significant contrast. A reasonable observer would conclude these differences will be telling for both groups.

I believe you will enjoy the points of view in *Cloud Sourcing for the Corporation* and hope that this work points the way to the future for you and your company. Good luck on your cloud sourcing journey.

Chapter 1

DESTRUCTION AND CREATION

—

Creates Opportunity for Buyers of IT Services

BUSINESSES and their underlying computer support networks are undergoing a radical shift driven by an amazing range of low-cost cloud services. This radical destruction of the old and creation of a new market provides opportunities for both buyers and providers of cloud services. Cloud sourcing, which is rapidly merging with and replacing legacy outsourcing, today presents a clear trend for companies, ranging from the world's largest corporations to small and medium-sized businesses.

THE END OF AN ERA

Today's business team has grown accustomed to leveraging highly-talented, low-cost resources around the world on a seamless basis. This era, which began in the mid-nineties, resulted in the creation of an amazing wave of service companies - mostly headquartered in India. Companies were built on low-cost, highly-skilled IT and Business Process Outsourcing (BPO) talent and radical changes in telecommunications costs that enabled the remote support of clients. This led to the creation of massive new IT services providers such as Infosys and Tata, as well as hundreds more, all fiercely competing for the global IT and business outsourcing market. The original legacy providers that predate this era, such as HP and CSC have, during this same time period, moved significant parts of their own operations to India and other lower-cost countries. Over time, it became difficult to tell the difference between India and US based providers as both now have significant onshore and offshore operations with similar price points.

CREATION AND DESTRUCTION

Now, enter the Cloud Providers who are leveraging new, disruptive technologies to create fundamentally different industry economics. An analogy can be made to disruptive fiber and IP switching technology that drove a rapid drop in global telecom prices, and enabled the creation of the offshore IT services industry where low labor cost could be accessed. The new cloud business model, however, is the key. It is based not on low-cost offshore labor, but instead on maximizing the use of hardware and software to a point that was not possible just a few short years ago. Using virtualization and other automation, as well as extraordinary investments in new data centers to drive 80-100% utilization of computing resources 24 by 7, cloud providers are once again creating fundamentally different industry economics. The purpose of this book is not to decompose the cost drivers that are its underpinning, as the price points of the cloud providers alone make our point.

But naked price can be deceiving. These investments in infrastructure allow cloud providers to provision cloud services 30-75% cheaper than traditional service providers with a dizzying array of services and delivery models. In the rush to create new models,

Cloud Factoid

Google reportedly invested more than $450 million in their Chicago area datacenter to support clients with low cost, automated cloud services.

there is now significant variation in pricing methods. This, in turn, renders obsolete some of the traditional methods of comparing prices due to the incongruence of varying pricing mechanisms.

THE POWER SHIFT

This new cloud sourced world shifts the balance more evenly across the globe between low and higher cost labor locations. This may already be changing the competitive balance between Indian and US-based providers. This rebalancing is due to the location of the computer resources. Hardware and software utilization are the main drivers of low-cost in the cloud. Smart cloud providers have the ability to automatically provision IT services and infrastructure with almost no labor involved. So the low-cost labor advantage of

India is now less important. We will talk more, in this and succeeding chapters, about those underlying keys and drivers of cloud pricing and the gold rush mentality around the creation of new capabilities and services.

HOW WILL THIS IMPACT YOUR BUSINESS?

Cloud services delivered at a materially lower price offer the business a chance to solve problems that previously may not have been addressable or to materially change how they do business. Some of the promises of cloud sourcing and cloud adoption are:

- Faster time to innovation as business can afford to provide seamless tools to a wider group of team members and get the whole team engaged in defining questions and solving problems;

- Access to a dizzying array of interesting business applications for your business that can be rapidly deployed and discarded if not helpful with growing your business;

- A material change in the cost of technology that will change the way we use and employ IT;

- The ability to adopt greater functionality with fewer headaches for users, as many changes will occur remotely, lowering the number of issues for all;

- Better security than most companies have on their own, because properly provided cloud services are encrypted when the data leaves your building, stored in an encrypted state and returned to you that way;

- The ability to rapidly ramp up resources to support your business with better dashboards which allow you to marshal resources at the click of a mouse;

- Lowering the capital requirements for a host of services which will allow you to redirect budget to the highest value activities for your enterprise; and

- Rapid provisioning of functional (e.g., marketing or engineering) applications provide businesses with more potential ability to quickly respond to changing needs.

EVALUATING THE RIGHT PROVIDER FOR YOUR COMPANY?

How do you evaluate which provider is right for your company? Business users are hungry for the new capabilities, promised by the emerging cloud providers, which appear to have a very low cost of entry. While many are arguing over the definitions of what cloud is, the need for the business to change, morph with the aim of growing revenue and control cost while delivering greater value to stakeholders, is relentlessly expanding the number of cloud vendors that need to be evaluated. The cloud sourcing market and the providers continue to evolve and change their cloud service offering with the intent of capturing market share and getting their clients to buy into *their* view of cloud. The bet is it will be sticky and hard to switch once their client has made the decision.

For some vendors, their cloud offering is but a thin veil over legacy capabilities while they build out true automated cloud infrastructures. An evaluation of the websites and presentations of some of these cloud vendors showed extensive depth in the definition of cloud services but very little about their own cloud offering or skills. This is often because the provider talking about cloud either has little substance to differentiate themselves in the cloud, or has a very small (but valuable) role to play as a tool or point software-as-a-service (SaaS) solution. As such, the provider has only one service dimension (dog) to sell and is forced to convince prospects that it is the best dog in the world for them, or he does not make the sale.

CLOUD COMPUTING DEFINITION

An evaluation of your choices in cloud sourcing begs to define the core question of what a true cloud sourcing solution is and how to give a broad enough definition to get your team looking in the right direction. Our analysis of the cloud nets a single definition of what it takes to be a true cloud player.

Cloud Computing Definition:

Cloud Computing provides on-demand network access to a shared pool of configurable computing resources that is rapidly provisioned and released with minimum client or provider interaction. This cloud model promotes availability and is composed of five essential characteristics and four solution types. (Refer to chapter 8 for more detail).

Many providers have the ability to provide services with minimal human interaction, but others are structured to require a people-based interface. This is regardless of whether the solution is based on a private, public or hybrid cloud environment. This human interaction may or may not be important to you, but it is an important distinction when filtering for cloud provider capabilities.

WHERE ARE COMPANIES AT IN THEIR JOURNEY TO THE CLOUD?

If you have not yet built out a cloud sourcing plan, you are in the majority. Few companies have developed a strategic plan for the cloud primarily because they are caught in an ever-increasing loop of gathering information on providers. Over 94% have not made any serious move with 53% still investigating the opportunities.

What steps have you taken towards engaging cloud services?

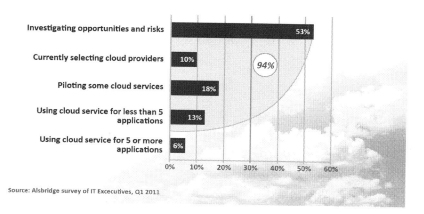

Source: Alsbridge survey of IT Excecutives, Q1 2011

DRIVERS FOR CLOUD

When executives were asked what their drivers were for investing or implementing cloud services, the leading drivers were reduction in operating costs, improving capacity to meet business demands and enhancing agility. The pattern suggests that while cost is a clear leader, driving business value and improving service levels were the widest area of interest.

What are your current business drivers for investigating implementing cloud services?

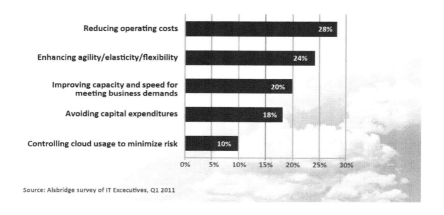

Source: Alsbridge survey of IT Exceuctives, Q1 2011

COMPARISON OF CLOUD PRICING
VS. TRADITIONAL OUTSOURCING

Our survey of Fortune 1000 executives and our initial experience indicates that cloud sourcing needs to be carefully evaluated in a revolutionary way that is counter to the traditional methods of evaluating the outsourcing market. Pricing that is publicly available from cloud providers will leave you with the promise that buying the services you need will be extremely cheap. Their prices can lead you to focus on the $.01 hour CPU pricing for a simple compute capability. For this amazingly cheap unit price to be realized, you have to prove your business case in advance. Our experience indicates the best way to prove the business case is to use a series of disciplined pilot programs for proof of concept and to provide inputs to your business case.

Data Points	Provider Comparative			
Components	Traditional	Joyent	Rackspace	Amazon
Base Machine	Varies	1 CPU Bursting to 8 CPU's RAM: 1 GB	CPU Min.Guarantee RAM: 1 GB	CPU capacity of a 1.0-1.2 GHz 2007 Opteron or 2007 Xeon processor
Base Machine Price Per month	LINUX: ($475) WINDOWS: $326	LINUX: $65 WINDOWS: Not supported at this price point	LINUX: $43.80 WINDOWS: $58.40	LINUX: $61.20 WINDOWS: $86.00
SLA	99.95%	100%	100%	99.95%
Instant Storage	Varies	15 GB Included in Base Machine Price	40 GB Included in Base Machine Price	160 GB (Non S3) Included in Base Machine Price
Real Storage @ 10TB	$.79 per GB Data transfer included	$.35 per GB 10 TB of data transfer Included	10 TB Data Transfer: 1TB/Out - 1TB/In 1M requests per month	Total for 10 TB @ S3 Price Data Transfer: 1TB/Out - 1TB/In 1M requests per month
Base Storage Price Per month	$7,900	$3,500	$1,856	$2,816
SLA	99.99%	99.999%	100%	99.9%

Source: PROBENCHMARK

This fast-moving market requires you to think about the problem differently and to understand that the cloud providers have very conflicting value propositions of how they will impact your business. You must first develop your own view, and then plan in an unbiased way regarding the possible as well as thinking about the contract and value proposition differently.

The question for the corporate buyer remains figuring out which provider will be right for his or her business and future. Many cloud providers are new and could be very helpful to your business. Should you consider that your historical outsourcing provider, who is now beginning to deliver cloud services, knows how to work closely with your management team and may be better for your needs? The right control and partner choice is a careful decision as you could inadvertently and unintentionally make decisions that lead to an unfortunate business design and architecture that achieves your goals by chance not plan. To help you think through the right cloud capability and provider, we have spent hundreds of man-hours developing a revolutionary way of looking at the range of providers available for your company. We hope you will enjoy what we call the Cloud Sourcing 100.

In the chapters that follow, we have gathered the seven keys to the cloud from our senior Alsbridge practitioners and clients who are on the front lines of enterprise cloud sourcing. Included are perspectives on the technology, the current market and provider community, the contracting and delivery implications and, most importantly, the journey today's CIO needs to take to begin mobilizing their enterprise to take advantage of cloud computing's potential to change today's business. It will be exciting times for all.

Chapter 2

CLOUD JOURNEY

The New Gold Rush

AS WESTWARD cloud expansion takes shape, many brave pioneers are starting on their journey toward the cloud. To safely and successfully make the voyage, these pioneers must carefully plan each step, anticipating the unexpected along the way. Cloud sourcing pioneers must make several important decisions. Are they even qualified to make the journey? What route should they take? What provisions are necessary to make the trip that far west? And finally, why is the journey to the cloud more important than the destination?

Cloud Computing is like Colorado in the 1800's. All people are out for themselves, and they will either strike it rich or die trying. Providers are already jockeying for a position in defining the standards of cloud computing while new cloud offerings and companies are popping up everywhere at a rapidly growing pace. To further exacerbate the ambiguity of cloud offerings, there seems to be a pioneer mentality to pave the trail for other features as well. Some vendors now sell a buggy or a saddle horse to get to the action quicker. But, the trail has its perils. In addition to dodging bandits (pop-up cloud providers), there are no maps, no GPS, and definitely no clear direction except 'Cloudward' - the trail dust has yet to settle.

It was an early morn' —brisk at best, cold at least. The sky was clear...not a cloud in sight. Jingle bobs parading down the planks and the pistols flashing where as big as the dreams of those who wore them. They all sat in the saloon, talking about what they were going to do when they strike it rich while others were just passing by, trying to get a feel or inclination to see if it was worth going West themselves. The smell of new paint on the walls, plush chairs, new set of cards at each table, and the chipper dealer with the twinkle in his eye gave off a good vibe and to those that wanted it -hope.

GATHERING YOUR SUPPLIES: SO MANY TO CHOOSE FROM

Cloud sourcing pioneers must make many decisions. Where is the best place to buy my supplies? Who will sell me what I need based on the conditions I am going to face and, ideally, has seen it all before? These are just some of the thoughts many CIO's and executives are facing daily when deciding if cloud is the right choice. There is some assistance however: 'trail guides' if you will. One of these is a company called CloudSleuth, which offers buyers an opportunity to look into how common applications are performing globally across various cloud providers using Gomez Technology. It quickly allows you to see firsthand what the response times are between these major Cloud Providers as well as measures content delivery networks and other various web analytics with simple applications. This is great if you are already on the trail (kind of like a weather forecast), but for those that are not even sure what trail they are going to take, it does not help much.

Another place someone may look is SpotCloud. SpotCloud has become this quasi buyer and seller exchange market for computer power. The mechanism is based on current market conditions, user demand, and level of resource utilization where sellers of excess computer capacity can sell it to prospective buyers. This is a broker site for cloud computing, and does very little for those that are not sure what they need to buy yet. As you can see, there are at least some attempts to help cloud pioneers understand the market, but little to actually help them enter the market.

Finding the right provider to purchase supplies from is not easy - there are many choices. What is typically needed in these cases is a 'trail boss' not a 'trail guide,' someone who has made this journey many times before and offers suitable advice based on your requirements and capabilities while offering a repeatable methodology and process that has assisted others who have traveled before you.

SOUTH BY SOUTHWEST, WEST, OR NORTH BY NORTHWEST: FINDING THE RIGHT ROUTE

One thing a trail boss will assist you with is assessing your capabilities and telling you what is possible based on those capabilities before you start shopping for supplies. Sometimes the hardest part of the job for a trail boss is telling someone they are not strong enough to make the journey. This could be due to infrastructure dependencies not available in the cloud, complexity of applications, or an organizational readiness that is just not up to par - and yes, it's common.

A Cloud Alignment Workshop (CAW) puts the reality in perspective quicker than Doc Holiday could skin his own smoke wagon! In this collaborative workshop, you review various case studies from people who have successfully made the journey already and examine how the providers helped them, review infrastructure and application maturity assessments conducted on your own organization, and explore how the capabilities of the providers in the market stack up against your own requirements and capabilities. If there are components that qualify, then it is a matter of setting a direction. Remember, this is just a qualification – the journey has yet to be planned.

The trail boss kicked the wheels on the buggy, pried each box open and checked the supplies with intensity that could only come from someone like that –his experience radiated in this moment. The client knew they had a good one. He untied his blue bandanna and wiped the sweat from his brow slowly stuffing it into his left back pocket and as he jumped down off the buggy; he looked at the client and said, "You're ready now." Gasps could be heard from the naysayers in the crowd that had gathered. It was determined, the route was selected.

IS THE ROUTE I WANT TO TAKE THE BEST ONE?
(IAAS AND PAAS COMPARED)

There is no such thing as the 'best route' in cloud computing. It is a matter of aligning your capabilities and requirements with a set of provider characteristics and delivery models that are either a match or not.

A provider properly matched to your requirements can be the difference between success and failure. Each provider is unique in their own right and no one provider can meet everyone's needs – although some will lead you into believing so. In the case of Infrastructure as a Service (IaaS), take a look at Amazon and Joyent.

Amazon AWS EC2 (at a basic level) is a replication of your infrastructure in the cloud. Joyent offers the same service claims but their technology stack is better designed for high intensity applications. What does this all mean? It means one thing; the language in itself is worth asking your trail boss about. Joyent is a great fit for some (like gaming and social media companies), whereas Amazon may be better suited for others who have more common requirements. It's the small things that matter here.

IaaS is an easy one to get once you are past the 'what is it I am getting for the money I am spending' part. Each provider has their own definition of compute. As an example, Rackspace offers a 512MB RAM setup whereas Joyent and AWS do not. It is worth asking your trail boss about before buying – after all you are only wanting to pay for what you need right? Another bullet to watch out for on the trail is the constant touting of HIPPA, PCI/DSS, and SAS70 compliance by some cloud providers. Just because someone made them compliant does not mean your auditor will see them the same way. Your trail boss should guide you gently down this path as well.

Platform as a Service (PaaS) is not easy to get and some people often mix PaaS and IaaS just to keep the conversation simple. But the reality is these are two very distinct components within cloud. A PaaS provider can sometimes offer both IaaS and PaaS. When you intend to buy a platform that an application runs on (often called a technology stack) then you are entering the PaaS realm. PaaS can be good and bad. Take the case with Microsoft Azure.

It is a .NET environment; this skill set is a commodity in the IT community and easy to find. But, should you choose to move to, let's say, Google App Engine, the preferred language is Java and Python. As you can see, it is important to plan these little things beforehand. Once IaaS and PaaS solutions are standardized, planning becomes easier.

It was getting dark. The constant sound of the wheels grinding over the rocks reminded the client of when she was a child in school, the scratching of fingernails on the black board by the other kids – she never quite cared for that. The ruts in the trail where many have set out before often edged the cliff. As she looked over the edge, she saw an upside down wagon that appeared to have rolled down the mountain and crashed violently at the bottom. As she looked closer she asked, "What caused that?" He said, "They chose the wrong trail boss." They all sighed in relief and were quickly asleep.

FOCUS ON THE JOURNEY NOT THE DESTINATION:
WHAT WE LEARN FROM CLOUD

Dawn approached faster than any others in recent memory. The sun was breaking through the clouds — most stood there facing it. The chill from the night was gone now; the warmth of the sun was better than the camp fire. There were others as well, as they slowly scanned the hillside they could make out more camps with each passing moment. Miners, stores, assay offices...it was industry at its finest. They had arrived! It was as if the entire journey only happened yesterday although it was the toughest journey any of them had ever made.

Advances in cloud computing are occurring at such a rapid rate that it is almost impossible for any one person to keep up while holding a day job at the same time.

However, the one thing cloud is teaching us is how dependent we have become on our current platforms, infrastructure, and various other IT systems and how much of an impediment these things can be to change. The reality is that some will never be able to take full advantage of cloud computing because the dependencies in their own infrastructure outweigh the benefits. Transformation into cloud computing will require a journey that is planned, mapped, and realistic. To enter this environment on your own without the proper research or help you are literally rolling the dice, crossing your fingers, betting on red —whatever analogy you choose. The last thing you want to do is to sign up for services before knowing what you need. To put it another way, hitting the trail without any planning, supplies, or trail boss.

Now that you have a better appreciation of the hazards that may lay ahead, the next chapter details the factors that an experienced "trail boss" would consider in developing a company's unique cloud roadmap.

Chapter 3

CLOUD ROADMAP

—

Creating Your Unique Cloud Roadmap

NO DOUBT you are seeing more and more information about the value of moving IT services "into the cloud." But do you know exactly what that means? Does anyone? Before moving into the world of cloud computing you must first gain a holistic view of your current IT environment and future business objectives. This way you will have a better understanding of the current realities and future possibilities available to your organization in the cloud. Your decision process should look much like a typical sourcing strategy process, with some special handling of cloud-related issues. Remember: one size does not fit all when it comes to determining your unique cloud sourcing strategy.

CLOUD SERVICE CATEGORIES

So what exactly is the cloud, and why is it receiving so much attention these days? There are many definitions and explanations being floated around, but one of the better ones that we use at Alsbridge comes from the National Institute of Standards and Technology (NIST) which states that cloud is, "A model for enabling convenient, on-demand network access to a shared pool of configurable computing resources (e.g., networks, servers, storage, applications, and services) that can be rapidly provisioned and released with minimal management effort or service provider interaction." Cloud services are categorized according to the purpose and intent of the services provided.

Typically these categories are classified as:

- IaaS (Infrastructure as a Service) - Virtualized servers, storage and network capability;

- PaaS (Platform as a Service) - Operating systems, development platforms and middleware;

- SaaS (Software as a Service) - Applications delivered over the internet (previously referred to as ASP - Applications Services Providers); and

- BPaaS (Business Process as a Service) – Evolving category of outsourced business process outsourcing (BPO) where discrete services are delivered via the cloud.

Furthermore, cloud services can also be classified as:

- Public cloud - Services made available to everyone over the internet;

- Private cloud - Privately owned with services made available behind a firewall to a restricted set of users; and

- Hybrid cloud - Interoperable combination of public and private clouds.

Depending on your specific IT environment and business objectives, certain portions of your current environment may be candidates to move into some combination of cloud computing environments. Some typical candidates for movement into the cloud include email (such as Gmail), customer relationship management (such as Salesforce.com), and specific applications that generate wide variances in server utilization, and result in heavily underutilized infrastructure investments.

CLOUD PROVISIONING STRATEGY

Developing and executing an IT sourcing strategy can be a daunting task. With the advent of cloud computing, the number of vendors and offerings to consider is increasing, which makes it even more important that you follow a defined, structured and proven process to evaluate your options and alternatives.

While it is easier in most cases to contract for specific cloud services (in some cases as easy as agreeing to the terms and conditions posted online by checking a couple of boxes), cloud computing is actually driving additional complexity into the IT and vendor management space due to multiple providers each supplying niche pieces of your overall infrastructure. The challenge is developing an overall sourcing strategy which allows you to bring each type of IT service into the overall IT environment, ensuring each piece functions as needed to seamlessly support your business.

In general, the process you use to develop your IT sourcing strategy, should address a key set of questions including:

- WHO will you outsource to?

- WHAT are you considering sourcing?

- WHEN are you considering sourcing?

- WHERE will services be delivered from?

- HOW will you transition services and manage the contract(s) and provider(s)?

- WHY are you considering sourcing certain services to a third-party provider?

These same questions also can be used to define the subset of your IT activities that may make sense to move to the cloud:

- WHO will you outsource to? In the cloud computing world, there are a multitude of vendors, many of whom are focused on one particular type of cloud service (IaaS, PaaS, or SaaS). Also, traditional outsourcing providers such as IBM, HP and Accenture are starting to build a stronger presence in the cloud services market.

- WHAT are you considering moving to the cloud? You will need to assess each application and infrastructure component of your portfolio to determine what may make sense to move. Consider the risks associated with transition, ongoing operations, security, regulatory compliance, and audit compliance.

- WHEN are you considering sourcing? The good news is that cloud services, in general, are easy to stand up and require less lead time than moving to more traditional environments.

- WHERE will services be delivered from? While the theory of cloud computing should make it location independent, the reality is that many clients still have a strong vested interest in understanding where services are provided from. Concerns may be related to performance, business continuity, government regulations, or a whole host of other reasons. Make sure you understand any location constraints you have prior to considering specific vendors.

- HOW will you transition services and manage the contract(s) and provider(s)? The bulk of transition efforts to the cloud will consist of converting and moving data into the new environment and the organizational changes required to train users on how to fulfill business processes using new technical solutions. While managing an individual cloud service agreement is normally simpler than managing a traditional outsourcing contract, your challenge will be how to manage multiple contracts, handoffs, and end-to-end service levels across a multitude of providers and technical environments.

- WHY are you considering the cloud? As you think about the potential savings, make sure you also think about additional cost and complexity you must build into your current organization to manage those services and to manage the integration of these services into your environment, such as licensing fees for software or decreases in current support models.

As you consider moving services to the cloud, keep in mind that you must offset the hype and sales promises bombarding you with a common sense, business-oriented approach that leads you to the best sourcing strategy for your organization. By analyzing your current environment, assessing different provider and technology options, defining alternatives, and building business cases based on a full understanding of the cost involved, you will develop a rational sourcing strategy that meets your current and future business objectives. When options and alternatives are unclear, consider piloting cloud services, as it can be done at relatively low-cost and low risk, as well as free. Engage a sourcing advisor to help you and your team to consider your unique circumstances and create a roadmap to sourcing the right services to the right providers and the right price and terms.

The next chapter highlights pricing trends in the cloud marketplace that you should be aware of while considering your cloud strategy.

Chapter 4

CLOUD PRICING COMPARISON

———

Pilots and Prototypes Point the Way

THIS CHAPTER briefly reviews the pricing structure of today's cloud market and its implications for adopting a cloud sourcing strategy.

Infrastructure as a service (IaaS) prices are readily available from most providers through their websites and follow similar constructs. Paas and SaaS pricing structures are more difficult and less meaningful to compare versus IaaS, since they are often based on unique differences in offerings as well as the value perceived by the buyer for the service in question.

Cloud server basic pricing across IaaS providers is typically based on a combination of server type (Linux, Windows, etc.), CPU speed, allocated processor memory, hard disk space and I/O speed. Each of these price drivers has a range of standard values that can be selected to configure the customer's server, resulting in a price measured in cents per hour. Of course, a variety of additional, value-added services can be added at additional costs. Enterprise buyers also need to consider additional factors including data compression, data transfer, storage, and latency variables when defining their specific requirements and server configurations.

The diagram on the following page illustrates the range of cloud server costs for a group of current IaaS providers for one of the primary server price drivers, Random Access Memory (RAM). RAM is becoming the de-facto pricing mechanism for a majority of IaaS providers, though a few offer non-mainstream price mechanisms as well with equal offerings. What can we conclude from this diagram? At the low end of server performance, pricing is quite similar. In addition, the straight shape of the lines indicates a consistent price per gigabyte for each provider as a customer is presented with higher performing servers (i.e., more gigs in the server). The key question is why are the slopes of the lines so different resulting in nearly a 3X difference between the lowest and highest priced servers at the highest performance range depicted (16 GB).

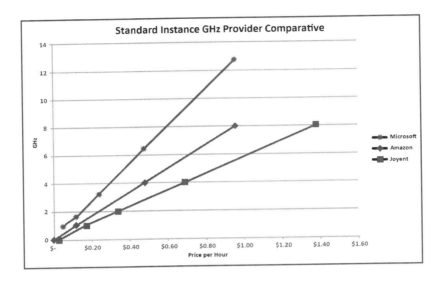

The answer is unclear because it is unlikely that the servers in question are truly commodities, that is, "all other things being equal" does not apply here, and in this there are important lessons. First, it is important to simply acknowledge that there can be a substantial range in costs depending on the provider/solution selected, so selection should be done carefully. Second, a clear understanding of what is dissimilar among solutions needs to be gained. These dissimilarities must be compared carefully to a well-documented set of related customer requirements in order to obtain the right provider fit and the greatest value for the customer.

However, these cloud customer requirements give rise to a new and unique challenge: it may well be the case that a customer enterprise does not have the tools to adequately measure application demand patterns and other characteristics necessary to create the "optimal" fit for a cloud-based solution. Cloud providers, fortunately, base their economic model on the ability to measure demand/usage drivers in great detail in order to be able to allocate their infrastructure across customers and create the high utilization that drives their economic efficiency.

Since clients do not typically have sufficient depth of demand data, running a pilot program with an initial provider of an application of interest is a great way to get a deeper understanding of the application demand profile. Fortunately, cloud technology has the added benefit of rapid provisioning with minimal set up costs so a pilot can be easily created. Permanent, optimal provider selection can be revisited after the pilot with the appropriate data. An entirely different provider may be selected based on the combination of application demand patterns and variances in provider pricing schemes putting companies on the right best-value curve in the diagram above.

Chapter 5

CLOUD CONTRACTS

———

Cloud Sourcing Contracts in an Outsourcing World

THE KEY to all successful outsourcing relationships (traditional, cloud, or hybrid) is the quality and flexibility of the commercial agreement between the customer and the provider. While cloud contracts have the same characteristics as any complex services agreement, the providers need to drive service standardization in both SLA and pricing are traded off for the promise of significantly lower price. While this can be extremely attractive, the buyer of cloud services needs to be careful to think through how to use the best of standardized cloud services while carefully evaluating the cloud provider's ability to overlay with custom services when truly needed.

In addition to establishing a good contract, whether it be a traditional or cloud services agreement, a key differentiator for success is the structure to manage the complete delivery of services as this establishes the approach for operating the contract, the interaction between the parties and, ultimately, the relationship. It is not the length or complexity of the contract that is the key to success.

The Alsbridge definition of a 'good' contract is one that is fair and sustainable between provider and customer, clearly defines the intended benefits and consequences (both good and bad) of the agreed relationship, and is based on a foundation of trust and partnership throughout the contracting process and beyond. A contract must stand the test of time and continue to meet both parties' expectations during the lifetime of the contract - not just at the start or at key milestones. This clearly applies to both cloud and traditional outsourcing contracts. The customer should receive the services and benefits they require, at the right time, while the provider operates a profitable enterprise and meets its own strategic objectives.

WHAT MAKES A SUCCESSFUL CONTRACT?

The focus for success is clearly on establishing the key principles for a high performing cloud sourcing relationship where many factors are kept in balance to achieve the net best outcome for the customer and provider. We mention provider here because it is in the best interest of the customer to obtain services from a financially and operationally healthy provider.

There are fundamentals to a contract document which include: scope, price, term, exit, liability, roles, and responsibilities… the list goes on. The key point is that it is not the contract which will break a deal; it's a breakdown in the customer/provider relationship. When both customer and provider agree on the principles for the deal and work together to meet a common set of expectations, the contract will evolve and continually reflect the business reality – what it needs is the mechanism to flex without creating administrative burden.

Establish the business objectives. It is critical to establish the business objectives. This is primarily to understand whether the case for outsourcing will support the business strategy and to ensure there is sufficient business case for change. Delivery of the business objectives will ultimately measure the success of the program; hence, it is critical to have these defined and documented.

Develop a complete cloud sourcing strategy. With the business objectives agreed to, the customer should develop the appropriate sourcing strategy to deliver the desired outcomes based on a proven set of cloud pilots that will establish the companies cloud strategy. The sourcing strategy will assess options including single or multi source agreements, assess current and future delivery models, and define provider selection criteria to ensure the best-fit provider(s) is matched to the customer and business objectives.

Many buyers (driven by mature procurement legislation) or policies, recognize the value of understanding the cultures of provider and customer organizations. This comes from in-depth, face-to-face discussions regarding the proposed scope and commercial models, and dialogues to test drive/pilot the proposed solutions. A collaborative, open and workshop-orientated approach enables both organizations to ultimately go into the deal with their

_, open and with full understanding of what is expected in order to deliver success.

To reduce risk, customers are turning to cloud sourcing advisors who have experience with a wide range of cloud providers and the ability to benchmark the correct price and service levels.

If customers do not use an external advisor, it is valuable to establish a level of impartiality and independence in the deal team and to ensure objectivity and challenge remain; it is also useful to help overcome any deep-rooted mindset(s) which may exist with key stakeholders.

To establish a fair and sustainable contract, be it traditional or cloud service oriented, the customer should provide a number of commitments to the provider:

- To provide quality time to the providers and commit to make decisions promptly;

- To be open and honest about the current costs;

- To recognize the economics of the deal and expect a fair price; and

- To be realistic about expectations of future improvements.

- In return, the provider should commit:

- To promise only what can be delivered;

- To provide a clear and complete picture of the services and pricing;

- To involve delivery people early in the process and focus on the solution rather than the deal;

- To align the economics of the deal and ensure a sustainable commercial model; and

- To recognize that the negotiation process takes time and will require give and take.

An open and honest contracting process involving customer and provider working in collaboration should result in a deal that both parties can make work.

ENSURING EXPECTATIONS ARE ALIGNED

Customers and providers should be crystal clear with each other from the outset about why they are entering the contract. When contracts fail , it is often because expectations have not been managed from both sides and the objectives of the deal fail to be realized. Managing expectations involves defining clear roles and responsibilities for all parties, but also relates to the principles of the deal. It may be the difference between a purely cost-cutting focus to growth orientated services. Instances of misalignment also arise in situations where the customer wants innovation and transformation, but the provider is not incented to achieve for transformation and continues to deliver the core service as usual.

Financial engineering is an area of the deal where expectations often fail. Advisory firms such as Alsbridge are often brought into distressed deals because customer circumstances changed part way through and the deal is felt to be no longer economically valid. Deal engineering (i.e., upfront cost savings, cash injections, paid for transition, etc.) ultimately needs to be paid for and has to be factored into any commercial changes. It is similar to a mortgage deal in that if you buy a fixed rate and want to switch to variable rate within the notice period, you will incur exit fees. The best contract is therefore one which has been firmly aligned to the customer's business expectations over the short and medium terms, has predictable and transparent costs and can be flexible within reasonable change parameters or breakpoints.

CLOUD SOURCING CONTRACTS: NOT IF, BUT WHEN

The commercial model and principles for contracting in today's outsourcing market will suit providers and those customers buying 'standard' outsourced IT services. These standard services will be less and less in demand as the market for cloud-based business services grows with customers demanding much more flexibility in how services are provisioned.

The traditional market for packaged software has peaked. Alsbridge predicts cloud-based delivery of applications will account for half of this market (1 million businesses already using Google Apps) within 5 years and that outsourcing of e-mail as a service will be routine within 2 years. Alsbridge's cloud sourcing survey

(November 2010) highlighted that 65% of respondents are either already using or are considering virtual infrastructure for applications. The commercial pressure and technical maturity of solutions have created robust infrastructure options offering a real alternative to standard infrastructure models. We expect to see rapid expansion and adoption over the next two years.

The commercial logic of the cloud is a major transformational journey that requires commitment from both the corporate-side and the provider community to deliver. It is a case of adapt or die; those slowest to market will be left behind.

For the corporate customer there is significant motivation to accelerate the process of transformation. In the current recession, corporations require a low-cost to change their businesses but seek significant benefits in terms of reduced operational costs and improved process flexibility. However, traditional providers have a significant motivation to resist and slow this transformation because their long established margins are at risk, and, in the virtual cloud-based world, there are minimal barriers for customers switching providers at the 'flick of a switch.'

In the short term, customers have much to gain at the expense of providers. Datacenters with less than 10% average utilization, unused licenses, slow ramp up for new services and applications are areas of redundancy that customers are looking to remove from their organizations. Contracting for these services in the cloud will become the new standard and will require a different set of contracting principles and a new basis for customer and provider relationships.

CONTRACTING IN THE CLOUD

Given its business to consumer roots, current contracting practice places much emphasis on the provider and potentially not enough on the corporate customer. Striking the optimum balance between ease of access and use and risk of service issues is important to get right.

With the potential benefits of highly automated services comes the providers' desire to automate and standardize the contracting processes in parallel (i.e. 'click to accept' terms and conditions); hence, attempts to renegotiate key provisions may well be resisted.

As with contracting for traditional ITO services, a detailed understanding of service provision and delivery is vital before entering into contractual commitments. Typical areas for focus during negotiations include:

- Service Level Agreements
- Service Credits
- Service Security
- Service Business Continuity
- Service Costing Formula
- Service Payment Provisions
- Data Ownership and Protection
- Legal and Regulatory Compliance
- IP Protection and Reward
- Software Licensing
- Software Escrow
- Choice of Law and Jurisdiction
- Third Party Rights
- Duration and Termination Rights
- Change of Control

The above is not an exhaustive list, but gives an idea of the complexity and detail required to enter into a cloud-based contract – "click to accept" is unlikely to be acceptable for many business situations. The difference with cloud is the commercial and business model within which it operates, the location of data, potential multi-tenancy agreements, on-demand hugely scalable and agile services (up and down), and interoperability are all significant issues which impact the 'consumer' style models compared to enterprise wide solutions.

It is important to think about that transparency of cost and the impact of being able to allocate cost back to individual business units within an enterprise. This is vital because that transparency brings accountability into the business user and puts the 'brakes' onto the demand model.

The principles for contracting in the cloud will have a step change on the outsourcing market and a significant level of work is being done to prepare the outsourcing market for the future of cloud-based outsourcing services.

We already know that cloud-based contracting will drive down the cost of software and infrastructure procurement and will allow much more transparent pricing as it will be on a pay by use basis. Other key areas of a typical traditional outsourcing contract may also be radically changed in the cloud world. For instance, intellectual property may be shared across organizations or removed from the equation altogether, and licensing for some aspects of the solution may be pooled or something that customers do not even need to take into consideration.

Cloud contracting offers easy switching between providers at the end of the contract or specific service requirement (e.g., after an IT project launch); customers may switch off or move to another provider. Providers may consider augmented contractual terms to curtail this movement. It may seem that cloud is tipping the balance entirely toward the customer in this regard, but customers have usually benefited from provider investment into the deal. How providers are rewarded for investment and discourage rate shopping is an area that will require careful balance when going through a robust contracting phase.

SUMMARY

While many customers will continue to contract for traditional outsourcing services as they have done in the past, we believe that many will shift a portion of the work that would have been otherwise outsourced to a variety of cloud providers. The need of cloud providers to achieve wider levels of standardization will create friction in contract definition, execution and management of contracts and management processes that will require leadership focus.

The incumbent ITO service provider that is beginning to address cloud requirements has the advantage of knowing the customer organization, the technical architecture, and the quick win cloud services it could implement. The friction that will make it difficult for them to effectively provide true cloud services will be the momentum of their high-cost, custom sales and delivery network that shudders at the click to buy of new age cloud providers. We believe the two versions of cloud provider delivery will coexist for a time period until the definitions of old and new contracting blur and merge into a new standard.

Now that some of the new challenges for cloud contracting have been discussed, it's time to gain a deeper understanding of the strategic use of the cloud delivery model.

Chapter 6

CLOUD DELIVERY

—

Five Steps to Unlocking the Secrets of Cloud Computing

THE EVOLUTION of service delivery models over time is significantly influenced by the level of business and technology innovation that occurs in the marketplace. Clients will need to change their sourcing strategies and providers will need continued investment in select emerging technologies in order to capture the promised benefits of new service delivery models.

Client organizations driven by economic and competitive forces continuously look to their IT service providers for improved economies of scale, productivity, and lower costs. In other words, they need their investment in information technology to help achieve real advantage in the markets they serve and to improve shareholder value.

Similarly, providers are driven to invest in new technologies to help the enterprise create growing interest in their product and service offerings, differentiate themselves from competitors, create barriers to entry in the markets they serve, and yield higher returns for their shareholders. Every so often, however, new technologies, processes, or service delivery concepts come together in the marketplace to fundamentally change the face of the industry. Facilities management did this in the late 1960s, as did the introduction of system integration in the 1970s. Managed services in the 1980s and the creation of the Internet in the 1990s further demonstrated how information technology service delivery models have changed.

Are we seeing yet another fundamental shift in today's established IT service delivery models? The emergence of cloud computing services, generally defined to include: Software-as-a-Service (SaaS), Software Platforms-as-a-Service (PaaS), and Infrastructure-as-a-Service (IaaS) appear to be edging into more established managed service delivery models. Certain providers claiming to deliver cloud services make it reasonable to conclude that some areas of cloud computing are relatively well-defined (SaaS) while others are still evolving (IaaS). Likewise, market analysts surveying the cloud services market space are indicating cloud services represent significant potential value to both providers and clients, but only if key business and technology barriers are overcome.

Despite a lack of clarity around whether cloud services are the next evolution of IT service delivery models, it is increasingly clear that organizations must re-evaluate their strategic sourcing alternatives. The sourcing strategy, future IT architecture plan, and contract terms must take into account the potential value of cloud computing. So, how do client enterprises determine whether cloud services make sense for them?

There are five tactical steps we suggest organizations consider as they try to answer this question:

1. Recognize that as technological innovation and business change drive the evolution of IT service delivery models, new sourcing strategies are required. Don't force-fit today's strategies and approaches to totally new sourcing alternatives.

2. Assess how conducive your organization's current IT infrastructure and platforms are to the introduction and roll-out of cloud computing in support of today's and tomorrow's business requirements.

3. View cloud computing as an expansion of delivery alternatives rather than a replacement of today's infrastructure and software environments. Most organizations simply have too much invested to abandon established enterprise infrastructure in favor of cloud services.

4. Initiate a dialogue with key stakeholders and key partners/providers to define key assumptions and critical operating parameters and determine how much flexibility exists in areas that may or may not be of strategic importance, including operational control and visibility, IT security, and performance-level management.

:velop a contingency plan that is aligned with your ,urcing strategy. Be prepared to act regardless of how much service delivery models change as a result of the introduction of cloud services.

In conclusion, emerging technologies force buyers and sellers of IT services to behave differently. Consequently, sourcing strategies must also change to meet new deployment alternatives. A reputable advisory services firm, such as Alsbridge, can help with experience, expertise, quantitative tools, and a view of the market to align sourcing strategies with evolving IT services.

Chapter 7

CLOUD CASE STUDY

MeadWestvaco's Approach to Using Google Apps

MEADWESTVACO (MWV) is a global leader in providing packaging and packaging solutions for their customers and has been a pioneer in the industry for more than 150 years. MWV has 19,000 employees that operate facilities in 30 countries.

Over the past several years, MWV has expanded through a number of strategic acquisitions which, among other things, has resulted in a proliferation of disparate email systems (Lotus Notes, Microsoft Exchange, etc.) used to support their global operations. Until recently, MWV has managed these systems in a traditional on-premise manner. MWV, like most large corporations, still ran their own email systems, including mail servers, gateways, client access servers, internal routing servers, public folders, email filtering, mailbox storage, and archiving, and were supported by a team of internal and external specialists.

The challenge for MWV was to engineer a highly-efficient, secure, consistent, global messaging environment that would better support their corporate "One MWV" business strategy. The email consolidation project was chartered with specific objectives, including to:

- Replace the ten existing messaging systems;

- Standardize email and collaboration tools globally;

- Reduce operating costs;

- Move from fixed-cost pricing to a scalable, variable-cost pricing structure; and

- Enable the rapid integration of new business acquisitions.

After considering key economic and non-economic factors (e.g., features, updates, usability), MWV decided to focus their evaluation on hosted cloud-based solutions. Cloud-based solutions are defined as systems that are run by someone else and operate on someone else's datacenter. They realized the success of their project would likely have broader implications regarding how they would support their other IT services in the future.

As their evaluation progressed, they became more impressed with the potential of cloud-based solutions in general, and decided to use the email consolidation project as a means of strengthening their understanding and knowledge of implementing and introducing cloud services into their corporate environment.

WHY GOOGLE APPS?

Security was a major concern for MWV. It was important for MWV to choose an email cloud service provider that had a proven track record for delivering secure, reliable and cost effective services. After considering multiple potential providers, Google Apps became a clear stand-out for the following reasons:

- Constructs its own hardware, server software, and operating systems for security;

- 140 million email users (Nielsen Research/ITU data; MWV analysis);

- SAS 70 Type II, EU Safe Harbor certifications; and

- Prime contractor for Department of Defense (DoD), National Security Agency (NSA), Central Intelligence Agency (CIA), Department of Homeland Security (DHS) IT "Web 2.0" projects

IMPLEMENTATION PILOT

To more clearly identify and address the potential risks associated with moving their email services outside the firewall, MWV designed an implementation approach that enabled them to learn and progress at a steady pace. This well-controlled cadence allowed them to surface any issues and address them in a highly-structured manner. MWV planned a series of pilots as the foundation for their implementation plan. The pilots lasted for a four-month period from March through June of 2009, and were conducted in two phases. The first phase was limited to just five associates and was intended to be a general proof-of-concept. After successfully completing the initial pilot, they initiated a second pilot where the number of participants was expanded to 110 associates. This encompassed representatives from 30 countries and all businesses units. During this phase the project team diligently captured issues, applied workarounds and identified long-term fixes, when appropriate, and determined risk-mitigation actions.

To better understand the end-users' perspectives, MWV conducted surveys at both the midpoint and end of each pilot period. This enabled the project team to assess overall satisfaction of the service to support their evaluation.

PILOT LESSONS

The pilot provided the IT team with valuable insight into both technical and non-technical aspects of the migration. That insight provided them with a better understanding of both the pros and cons, including those listed below:

PROS
- Users indicated a strong preference for Google email service (90%+ of users surveyed agreed or strongly agreed that the company should move to Google);

- Enabled migration to a single platform in a 6-9 month period;

- Provided a strong collaboration capability; and

- Accessed a constant stream of innovation and functionality improvements from Google.

CONS
- Lotus Notes and Google calendar synchronization was very difficult (required users to perform task manually); and

- Security issues required policy changes and training for end users.

During the pilot, MWV also gained insight into the full-suite of Google Apps services beyond email. With their improved understanding they determined the expanded suite of services were about 12-18 months away from being able to replace MS Office for their average user. This clearly provided them with a viable roadmap for future expansion of cloud-based services.

POLICIES REVISITED

Confident with lessons learned from the pilot, MWV began work on revising certain corporate policies that would better support their new operating approach and better ensure the security and protection of MWV data. Some of the areas revised included:

- **Document & Site Sharing.** Using Google Apps introduced more data searching capability and collaboration into the MWV environment. As such, there was a greater potential for sensitive or confidential data to be shared inadvertently. To mitigate this risk, users required training to promote their awareness of the need to protect sensitive information. There was also a need to limit document and site sharing to the MWV domain.

- **Email Retention.** The MWV 90-day email retention policy was not uniform globally and required the establishment of a standard that met the global business needs. The policy was changed to standardize on a one year email retention policy.

- **Email Archiving.** MWV users archived email beyond the requirements of the retention policy using automated features. The policy was changed to disable automated archiving and to require users to perform selective archiving consistent with the records retention policy.

- **Internet Access.** Email could now be accessed anywhere and anytime from a browser. New training was established to ensure business users understood browser security and logoff requirements.

BENEFITS

The following are benefits that MWV has realized with their first significant migration of services to the cloud:

- **Advanced data protection and security.** The security of the email system is enhanced as a result the expertise and investments made by Google in protecting the environment.

- **Allocation of IT professionals to more business-focused activities.** Moving services and equipment to outside providers has enabled the IT department to concentrate more of its efforts toward addressing their business customers' needs.

- **Immediate access to the latest software and configurations.** There were often delays and significant hassles with upgrading new software releases. End-users now have more rapid access to continuous introduction of new Google features.

- **Avoidance of capital expenditures.** Consolidating their disparate email solutions into an expanded on-premise solution would have required additional fixed-cost investments and added time to the implementation. MWV avoided capital expenditures with a variable-cost solution where they only pay for what they need.

- **Leveraging massive storage capabilities.** MWV can rapidly provision additional storage capacity as user needs dictate. This enables a more rapid integration of acquisitions.

MWV approached their venture into cloud computing by taking a well thought-out approach at a pace that enabled them to develop capabilities and confidence. As a result, they were able to achieve their project objectives and are better prepared to continue their journey in the cloud.

ADVISE TO THOSE CONTEMPLATING A CLOUD JOURNEY

- As you venture into the cloud, take slow steps to begin with; there is much more to learn than you may have imagined.

- Well-designed and managed pilots are an excellent tool for uncovering and addressing implementation challenges.

- Address one piece of the cloud as you develop your understanding of the services available to support your business.

- After you start your journey, you will be surprised how your views and perspectives of the cloud are likely to change.

- Understand corporate policies and how to integrate or change them; this will be a critical element to insuring proper integration.

Chapter 8
CLOUD SOURCING 100 PROVIDERS

———

Comparison of Capabilities

THIS CHAPTER introduces the *Alsbridge 100 Cloud Sourcing Providers* classification system, explains the system's origin and defines its structure in detail. These descriptions are followed by the individual profiles and classification of 100 of today's cloud service providers.

ORIGIN AND USE

The Alsbridge Cloud Provider Classification system is based on CIO responses to a series of structured questions and discussions of cloud buying criteria that took place during Alsbridge's Outsourcing Leadership Summit and CIO Round Table. The content on each provider is derived from Alsbridge's primary industry research including many in-depth interviews with provider representatives when appropriate. The system is *not* a ranking of the providers included, but, rather, it is a classification/categorization system intended as a guide for identifying the cloud solution type, provider type, degree of customer control, and unique features and characteristics of their respective delivery models and companies.

DEFINITION OF THE CLASSIFICATION SYSTEM

The classification system used in the individual provider profiles has a number of component parts each of which are explained in this chapter.

SETTING THE FOUNDATION: CLOUD SERVICES DEFINED

The industry has struggled to agree on a precise definition for cloud computing as it evolves, but we find the definition developed collaboratively by the National Institute of Standards and Technology (NIST) to be a perfectly adequate place to start. NIST defines cloud computing as, "a model for enabling ubiquitous, convenient, on-demand network access to a shared pool of configurable computing resources (e.g., networks, servers, storage, applications, and services) that can be rapidly provisioned and released with minimal management effort or service provider interaction." This cloud model is composed of five essential characteristics:

On-demand self-service. A consumer can unilaterally provision computing capabilities, such as server time and network storage, as needed without requiring human interaction with each service's provider.

Broad network access. Capabilities are available over the network and accessed through standard mechanisms that promote use by heterogeneous thin or thick client platforms (e.g., mobile phones, laptops, and PDAs).

Resource pooling. The provider's computing resources are pooled to serve multiple consumers using a multi-tenant model, with different physical and virtual resources dynamically assigned and reassigned according to consumer demand. There is a sense of location independence in that the customer generally has no control or knowledge over the exact location of the provided resources but may be able to specify location at a higher level of abstraction (e.g., country, state, or datacenter). Examples of resources include storage, processing, memory, network bandwidth, and virtual machines.

Rapid elasticity. Capabilities can be rapidly and elastically provisioned, in some cases automatically, to quickly scale out, and rapidly released to quickly scale in. To the consumer, the capabilities available for provisioning often appear to be unlimited and can be purchased in any quantity at any time.

Measured Service. Cloud systems automatically control and optimize resource use by leveraging a metering capability, typically through a pay-per-use business model, at some level of abstraction appropriate to the type of service (e.g., storage, processing, bandwidth, and active user accounts). Resource usage can be monitored, controlled, and reported, providing transparency for both the provider and consumer of the utilized service.

THE FOUR COMMON SOLUTION TYPES OF CLOUD SERVICES

With this general definition of cloud services in place, it is valuable to more deeply distinguish cloud services. To this end, the industry has adopted four service solution definitions whose use is broadening. These four solution types represent the first component of the classification system and are prominently indicated by boxes on the top of each profile page containing the first letter of each solution type.

The four solution types are:

- Infrastructure as a Service (IaaS);

- Platform as a Service (PaaS);

- Software as a Service (SaaS); and

- Business Process as a Service (BPaaS).

While all solution types conform to the general cloud definition above, each is very unique with respect to where it fits in a customer's IT and business architecture. The value in focusing on these distinctions is to enable buyers to focus on IT components being evaluated for transformation into the cloud, instead of having a selected provider dictate which services can be delivered resulting in limited options, technology lock-in, and the creation of high risk transactions.

INFRASTRUCTURE-AS-A-SERVICE (IAAS)

The first cloud solution type is Infrastructure-as-a-Service (IaaS) and can be defined as:

Cloud infrastructure made accessible to a customer such that the customer has the capability to provision processing, storage, networks, and other fundamental computing resources, and deploy and run arbitrary software, which can include operating systems and applications. The consumer does not manage or control the underlying cloud infrastructure but has control over operating systems, storage, deployed applications, and possibly limited control of select networking components (e.g., host firewalls).

IaaS Sub-Classifications

IaaS services represent a broad and active segment of the cloud market today, and, as such, further research and deeper classification of these services is helpful to customers investigating cloud services selection implementation. Within the individual provider profiles that follow, a table entitled "IaaS Sub-Classification" lists more in-depth characteristics of IaaS services (see diagram 1). These characteristics are explained below and each provider supporting them will have a check mark in the appropriate line of the IaaS sub-classification table. We first explore two important characteristics that distinguish the provider companies themselves and then detail the many available components of IaaS service offerings.

- **White Label Providers.** The name derives from the image of a white label on packaging which can be filled in with a reseller's trade dress. In this context, the resellers are infrastructure service providers who are selling access to cloud services that are owned and operated by "white-label" service providers -- companies that have chosen to make the capital expenditures in the technologies and architectures that enable cloud. Resellers of white label services may have access to unique market segments or installed customer bases in need of IaaS services. Resellers may sell IaaS to customers with different levels of unique value-add including different features and options.

  Because the infrastructure agreement between a reseller and its customer is typically managed through service levels and other traditional sourcing service agreement mechanisms, this concept is generally accepted since it shares the risk between the provider and the customer.

  Often the customer's decision to acquire services from a reseller versus directly from an IaaS provider becomes one of comfort with its own IT organization's current organizational compliance measures, reporting accuracy, and process efficiency. If satisfied with its internal practices, then a customer could buy directly from a white label IaaS provider. However, if a customer were struggling with making routine changes on a daily basis, then certain providers would not be recommended. Provider selection is not always purely a financial decision but sometimes is a function of the fit between buyer and provider maturity and capabilities. Keep in mind, there is nothing wrong with receiving services from a reseller. It is often purposeful and reseller's should disclose this information up front first. These types of providers exist for reasons that meet the needs of today's evolving cloud market.

- **Traditional Providers.** "Traditional" providers are today's large IT Services companies that are distinguished from emerging cloud entrants because of their large installed base of global IT infrastructure, global customer relationships, and breadth of IT services. Historically, it has been reasonably straightforward for these types of providers to adapt to technological advances such as Remote Infrastructure Management and other variances of virtualization as they were often early adopters of these technologies. Cloud adoption by these traditional providers has proved to be much more challenging as they must carefully manage both the cloud opportunity and its threat to their existing business model and services. As such, it is currently unclear what role these traditional service providers will play in delivering cloud in the future. Evolving strategies by some of the more notable companies have created higher than normal degrees of confusion to buyers of IT Services.

Traditional IT service providers who cannot offer a pay-as-you-go service today may follow a reseller path and contract with "purer" IaaS white label providers like Amazon Web Services or Rackspace.

As their cloud strategies solidify, traditional providers have the potential to bring a strong positive impact to enterprise customers by leveraging their existing maturity and strengths, such as best practices, ITIL process maturity, service management centers of excellence, governance models, change management processes, and integrated cloud offerings mixed with traditional services. There are indications that some of the traditional providers are moving in this direction already with their new offerings claiming cloud-like capabilities.

- **Servers.** Servers are a primary component of IaaS services and can be a distinguishing characteristic of an IaaS provider; that's because not all servers in the cloud are created equal. Qualities that distinguish cloud servers include server type (i.e., Linux, Windows, etc.), processing power, burst processing, RAM and instant storage, and their related pricing mechanisms. These varying degrees of technological sophistication have made "server shopping" somewhat difficult for the average buyer to dissect. Some providers will offer bursting capabilities free of charge in multi-tenant environments but may offer less instant storage, while others will allow bursting capabilities and massive amounts of storage. Another example of the differences is that some providers will also charge on a daily average for utilization, whereas others will charge for utilization on a monthly basis – both are suitable depending on the customer's objectives.

  These fundamental differences between provider server offerings demonstrate the need to understand customer requirements carefully before engaging a market full of diverse offerings.

- **Storage.** Cloud storage capability is also a core component of IaaS services. However, cloud storage solutions can be difficult to compare since provider storage offerings are often blended with other features. There are generally three types of cloud storage to consider: persistent, non-persistent storage, and local machine storage. The differences between persistent and non-persistent storage is a more technical discussion suited for data managers, however, each has its own uses, benefits, and drawbacks and you should refer to your data manager to better understand these requirements.

  Local machine storage (sometimes referred to as "inclusive") is typically tiered based on the configuration of the machine being provisioned. This type of storage is a fixed capacity and each customer configuration will require a certain amount of local storage. Once again, referring to the appropriate group in your organization to determine these requirements beforehand is recommended.

Common customer mistakes when purchasing storage are usually due to a lack of understanding of their own environments, or are due to poor data management processes. As a result, high-risk transition plans are created that often lack data risk mitigation commonly found in the traditional sourcing models. For customers who lack consistent data management, buying these storage services from a traditional provider may be a viable option.

Cloud storage solution pricing models are often based on gigabytes per month; however, differences in these models do occur. Incremental storage pricing options (buying blocks of storage) can result in purchasing more storage than is required and can be advantageous for those that understand their own storage requirements. However, for those that have unpredictable storage increases and decreases, a price per actual gigabyte would be more suited since buying incremental storage (storage blocks) may result in underutilized storage, defeating the purpose of going to the cloud all together. Because of these differences, a cloud storage strategy and roadmap compared to provider delivery capabilities is highly recommended.

- **Disaster Recovery.** Cloud itself is not a solution for disaster recovery but should be viewed as an enabler resulting from the data replication and storage capabilities of cloud providers. Most enterprise clients do not have this replication capability or lack sufficient capital to implement disaster recovery themselves. This is often the starting point for conversations that eventually turn to the topic of storage as previously described. In the event of a disaster, the services can be quickly reproduced based on the current stored data similar to traditional Recovery Point/Recovery Time Objectives (RTO, RPO) found in traditional disaster recovery agreements. Companies such as Iron Mountain offer disaster recovery as part of a broader storage offering including archiving and data retention consistent with regulatory compliance, and assist with bringing the services online in the event of a disaster as the customer and provider operate with an integrated process.

Some cloud providers however may not specialize in disaster recovery but offer similar storage capabilities. These providers may or may not provide the assistance needed in bringing services online in the event of a disaster, but they will have your data readily available. These providers should be investigated and their delivery role integrated into the broader enterprise disaster recovery plan.

- **Backup.** Backup services delivered via cloud providers is an ambiguous topic as backup is often associated with storage and depending on which provider you assess, the definition will vary. What is commonly associated with backup however is the data that resides on an end user device (e.g. laptop, desktop, mobile phone) that is sent to a provider data center where it is stored, and in the event of a device failure, the data can be retrieved and the device restored back to the pre-failure state.

- **Messaging.** There are generally three types of messaging providers available through cloud services. The first is common and is associated with web based messaging such as Gmail through Google. The second type is providers who will host the platform and perform the management functions for applications such as MS Exchange, and are often a reseller of the particular application. These types of providers may only offer this component as a standalone service (i.e. not part of a broader IaaS offering), and should be considered highly specialized niche players. The third type of messaging provider is an Independent Software Vendors or ISV. There are countless ISVs in the cloud market today, all of whom offer a piece of messaging software and often times sell the software as Software-as-a-Service (SaaS, explained later in this chapter). Depending on customer objectives, either one of the three common offerings can be well-suited. Some degree of caution should be used during the provider selection process as license transfer and other fees can apply.

- **Content Delivery Network.** Content Delivery Networks (CDN)s distribute web content to remote corners of the cloud to provide end users with faster (lower latency) access to that content. CDN's can be a critical component to the successful implementation of an enterprise's web architecture and overall system performance with significant potential impact on customer service and brand loyalty (e.g., media companies with frequent video access). However, CDN impact depends on the enterprise's unique need to leverage highly responsive web content and, therefore, may or may not be an important consideration when considering IaaS services.

  CDNs have been in existence for over a decade and are by definition cloud services. Enterprise customers can acquire CDN services directly – a CDN provider will place a node (web server) in a public or even a private network to assist with offsetting delays. Or, the need can be handled by the core IaaS provider. Instances of cloud infrastructure providers such as Rackspace forming partnerships with companies such as Akamai, who specialize in content delivery, are becoming commonplace. However, there are companies such as Voxel who offer a "triple play bundle" of infrastructure services, storage, and content delivery. Each CDN provider may have different enhancements and offerings that make them unique such as distribution and geographic coverage of edge devices, pricing mechanisms, and integration experience with cloud service providers.

- **Desktop as a Service (DaaS).** The ability of a user to login into a web portal from any machine and access desktop, software, server, email and all the other things normally associated with a personal machine can now be done in the cloud through a service designated as "Desktop-as-a-Service." The reality is that this has been available for some time, since a few select companies figured out how to deliver it years ago, but the "as-a-service" designation leverages the marketing power of today's cloud hype.

There are numerous DaaS offerings in the market that differ based on price and software breadth. Providers of DaaS generally fall into two categories: desktop specialists and broad-spectrum IaaS providers. Examples of specialists' services include one from icloud.com which offers a windows environment, advanced desktop features, 100GB of storage and backup, unlimited uploads, current patches and releases as well as anti-virus and other features for an economical monthly price. Another company, Cloudworks, offers a full suite of DaaS including Microsoft Visio and Project, 1GB of email, 5GB of storage and nightly backups. Additionally, Cloudworks offers and enhanced small business service including Quickbooks, CRM access and others with monthly pricing, blurring the line with SaaS based delivery models.

- **Hosting.** Hosting has become one of the most debated terms in the industry today. Before cloud technologies enabled hosting to take an enormous leap forward (indeed, many cloud providers evolved from hosting companies), the term represented a spectrum of services from "co-location" in which the customer supplies the servers while the provider supplies the power and facilities, to "hosting" in which the provider includes installs, racking and stacking, to "managed hosting" in which the provider manages everything except the application on the machine. A few years ago to say, "I want something hosted," was usually followed up by, "What type of hosting do you need?"

Today the industry is struggling with defining the difference between hosting and cloud computing. There have been several attempts at this, but most have been met with low-tolerance by purist in the community who need to ensure that the specificity of the offering qualifies in one category or another. What has been generally agreed upon is that "hosting" includes dedicated servers that are not "utility" (the core of a cloud delivery model). However, industry innovations are even causing that distinction to blur now. Companies such as Nephoscale and Voxel are now offering inexpensive dedicated private servers on a monthly basis that can be canceled at the end of the 30 days. For the purposes of our profile, cloud providers receive a "hosting" check if they offer dedicated servers in addition to more flexible cloud models, including short-term dedication models.

• **Information Technology-as-a-Service (ITaaS).** While customer empowerment and independence seem inherent in the definition of cloud computing, IaaS providers have recognized the market need to provide more traditional value-added IT services or managed services in conjunction with IaaS. This higher level service has been called "Information Technology-as-a-Service" (ITaaS) and is defined as IT services delivered on a pay-as-you-go basis, consisting of virtual private data centers or virtual servers in the cloud, combined with outsourced IT operations services including systems design, setup, operation optimization, security, and incident response. Companies such as Enki.co not only offer cloud-computing services in both the private and dedicated space, but also offer the ability to manage it for you. Traditional service providers are common in this space; however, for those that lack the pay-as-you-go model, the line blurs between ITaaS and traditional managed services. Each has pros and cons depending on the type of delivery required for the customer.

• **Management Software (S/W).** IaaS service offerings include a customer interface that provides software, dashboards or portals necessary for the customer to manage and monitor the acquired cloud infrastructure resources. However, additional infrastructure Management Software to manage IaaS resources ranging from server configuration, to application management, to data exposure control is widely available through independent software vendors (ISVs) and often delivered via SaaS, but can be traditional licensed software. Companies such as VMWare, RSA, and Oracle offer software that perform very specific functions and are common today in most environments not just limited to cloud. Often the software can be purchased and used in conjunction with IaaS provider offerings. This is the case with CohesiveFT VPN-Cubed Cloud Only solution, which is compatible with GoGrid, Flexiant, Terremark, and Amazon Web Services technology.

PLATFORM-AS-A-SERVICE (PAAS)

The second cloud solution type is Platform-as-a-Service (PaaS) and can be defined as:

A development platform that enables the customer to deploy onto the cloud infrastructure consumer-created or acquired applications created using programming languages and tools supported by the provider. The consumer does not manage or control the underlying cloud infrastructure including network, servers, operating systems, or storage, but has control over the deployed applications and possibly application hosting environment configurations.

Platform-as-a-Service (PaaS) is the lesser known of cloud delivery models. More detailed PaaS characteristics are mainly specific to development and testing, but also include other features that are too technical to discuss in the context of this book.

Many view PaaS as the ability for a developer to go into a virtual environment, load an application onto a server located elsewhere for testing and development, and pay for the use of that server only while utilized. PaaS has worked well for companies that are creating Software-as-a-Service products, and for enterprises that need to develop and test corporate applications being ported to the cloud. PaaS enables cloud development without requiring significant purchases, navigating a lengthy procurement process, or waiting for the IT department to install the shipped equipment. With PaaS developers are up and running in just minutes.

PaaS providers differ based on the supported programming languages and application themes. Some PaaS providers may only offer one supported language and platform for development, creating some degree of technology lock in. Selecting a PaaS provider requires a clear understanding of development requirements for the specific applications to be ported or developed. It would not be good to enter into Google Apps only to find out your developer requires vendor support for a .NET environment.

SOFTWARE-AS-A-SERVICE (SAAS)

The third cloud solution type is Software-as-a-Service (SaaS) and can be defined as:

Applications running on the provider's cloud infrastructure in which the applications are accessible from various client devices through a thin client interface such as a web browser. The customer does not manage or control the underlying cloud infrastructure including network, servers, operating systems, storage, or even individual application capabilities, with the possible exception of limited user-specific application configuration settings.

Software-as-a-Service (SaaS) is the entry point for many companies into cloud services and the number of offerings is staggering. Gmail, Adobe Flash, and various anti-virus applications are common examples of SaaS applications. Today nearly every vertical market has SaaS solutions being offered either by new entrants or by incumbents porting existing solutions. The distinguishing characteristic between SaaS providers, next to service offering, is their ability to deliver service on either a subscription or pay-as-you-go model. Both are common and equally effective.

BUSINESS PROCESS-AS-A-SERVICE (BPAAS)

The fourth cloud solution type is Business Process-as-a-Service (BPaaS). BPaaS is an emerging solution and relatively little is known about the impact it will have on traditional BPO offerings. It is believed by some that a standardized process will be difficult to sell as a service to enterprises due to unique requirements and that those that do so are subject to "process lock-in." Others, however, view the BPaaS environment as extremely beneficial with process offerings such as Medical Coding and Billing, Statistical Modeling and Regression Analysis, and Mortgage and Loan Processing. There are only a handful of players in this space such as WNS, IBM, and Genpact. The BPaaS market will be an interesting market to watch in the coming year as companies who laid-off employees during the recession scramble to acquire skill-sets and ramp volume to meet post-recession demand while maintaining a less-risky, variable cost structure – all the professed benefits of BPaaS.

KEY COMPARATORS: PROVIDER TYPE
AND CUSTOMER CONTROL

In our research of today's cloud providers two characteristics emerged as keys to quickly understanding basic but meaningful differences among this large array of service providers. They are "provider type" which classifies service breadth, and "customer control" which classifies the amount of solution control provided to the customer. Each is described below.

Provider Type
Provider Type classifies each provider based on the breadth and focus of services they offer to the marketplace:

- **E – Emergent.** These providers have a single focused solution but are relatively new to the market and do not yet unequivocally fit into one of the other categories. Some traditional outsourcing service providers can fall into this category as well due to limitations in delivery options (particularly cloud fundamentals like self-provisioning), leading to some cloud offering ambiguity.

- **C – Complementary.** These providers supply complementary technologies or services important to the four cloud solution types, but are not complete cloud solution or platform providers themselves, whose offerings are broader. Examples include providers of Content Delivery Networks, Cloud Management Software, as well as other Independent Software Vendor applications (e.g. risk management, provisioning, database management, etc.). Highly specialized or customizable components of cloud technology are included in this category.

- **F – Focused.** This classification is usually applied to providers who have limited (one to two) service offerings in total, or may have several offerings but only support a limited amount of operating systems and/or programming languages. Additionally, SaaS providers today are predominantly classified as "Focused" providers with single, specific horizontal or vertical offerings; however, some SaaS providers may offer additional infrastructure or development platform cloud services which would place them in a different category.

- **B – Broad.** These providers have three or more solutions, support multiple programming languages and deploy multiple operating systems. There are several service options such as server and storage configurations, pricing, and multiple platforms for SaaS development. Some offer white label infrastructure services will hold several compliance certifications (e.g., HIPPA and PCI/DSS).

- **I – Integrated.** These providers are able to offer enterprise customers cloud solutions and services that enable the customer to create a seamless, integrated architecture that includes traditional sourced infrastructure, private cloud, and public cloud solutions. Few cloud providers in the market today are classified as "Integrated." This represents an aspirational category for traditional providers as they begin to synthesize their traditional services, assets and delivery model with cloud technologies and business models in an effort to meet the demands and challenges of their enterprise customers with complex legacy architectures.

Customer Control

Customer control is used to identify the level of customer interaction and transparency within the cloud provider offering. Since each provider offers different types of services and technology, some control may or may not be necessary. Low Control, Medium Control, and High Control environments have different benefits and drawbacks depending on the specific needs of the customer.

- **High Control –** Customers of this type of service offering have the ability to manipulate the environment through a management tool or application. This may consist of root and administrator access privileges, and/or multiple operating system and programming language choices. These environments may or may not be supported entirely by the provider, or the provider may charge fees for premium support. Self-help and community portals are common in some instances, requiring skill sets that may or may not exist in the current customer organization.

- **Medium Control** – Customers of this type of service offering have a limited ability to manipulate or view the environment. Root access is not allowed and/or there is no management tool or application. The provider controls the root and administrator access privileges, manages the updates and patches to the operating systems, and supports the programming language(s). Control responsibility is shared between the provider and the customer. This eliminates the customer's need to retain or hire system administrators. This classification is common among traditional outsourcing providers who offer cloud services.

- **Low Control** – SaaS software that does not allow root access or offer a high amount of customization results in a low control classification. Updates, patches, and installs for the software are commonly controlled by the provider with little interaction required from the customer. Content Delivery Networks and other niche providers also receive this classification since little or no management is required from the customer.

ADDITIONAL PROFILE COMPONENT DESCRIPTIONS

The four primary solution designations and the sub-classification of IaaS services detailed above are critical to primary classifications of today's cloud providers. The Provider Type and Customer Control classifications are constructive for quickly facilitating a comparison of providers. However, many additional attributes are relevant to gain a deeper understanding of each individual provider. These are included in the provider profiles and are described below:

Provider Summary
The summary below serves as a description of the provider and their overall service offering as articulated by the company itself. This information was obtained from various websites, some provider interviews, and the Alsbridge provider database.

Year Founded

The year the company was founded.

Moody's Rating

The companies last known general credit rating as determined by Moody's based on publicly available information[1]:

- **Aaa** Obligations rated Aaa are judged to be of the highest quality, with minimal credit risk.

- **Aa** Obligations rated Aa are judged to be of high quality and are subject to very low credit risk.

- **A** Obligations rated A are considered upper-medium grade and are subject to low credit risk.

- **Baa** Obligations rated Baa are subject to moderate credit risk. They are considered medium grade and as such may possess certain speculative characteristics.

- **Ba** Obligations rated Ba are judged to have speculative elements and are subject to substantial credit risk.

- **B** Obligations rated B are considered speculative and are subject to high credit risk.

- **Caa** Obligations rated Caa are judged to be of poor standing and are subject to very high credit risk.

- **Ca** Obligations rated Ca are highly speculative and are likely in, or very near, default, with some prospect of recovery of principal and interest.

- **C** Obligations rated C are the lowest-rated class of bonds and are typically in default, with little prospect for recovery of principal or interest.

Pricing

Pricing is marked Y (yes) if Alsbridge has pricing information for the provider's services, N (no) if not.

Sales Channels

---

[1]http://www.moodys.com/sites/products/AboutMoodysRatingsAttachments/MoodysRatingsSymbolsand%20Definitions.pdf

Sales Channels displays publically available information about partnerships, resellers of the services, and traditional sourcing provider relationships.

Clients

Client information that is publically available through either customer stories or client listings available on the provider website at the time of research.

Termination Notification (Days)

Termination Notification represents the amount of time required in the event a buyer cancels the service voluntarily. A zero (0) represents immediate termination and Negotiable is represented by (NG).

Dedicated Account Management

The provider will assign at least one representative of the company as a single point of contact to handle all matters and issues pertaining to the services in most instances.

Live Customer Support

The provider currently has either a contact number or live chat feature for end-user support services.

Guaranteed Network Availability

The provider either offers minimum network availability through varying degrees of custom service-levels or offers a standard uptime metric that is applicable to customers.

Root/Administrator Access

The provider extends root access on Linux based operating systems or administrator access for Windows operating systems.

Portal Support

The provider offers a web based URL either through a secured portal or through a public portal to a knowledgebase containing service information, community development, questions and answers as examples.

Regulation and Compliance

Regulation and compliance will list the provider capabilities to deliver compliant solutions. These types of compliance are commonly associated with SAS70 TYPEII and PCI/DSS delivery capabilities as well as those that have actually obtained the certifications themselves.

Programming Languages Supported

Programming languages supported vary between providers. Due to the high frequency of change in provider delivery models, this list should be viewed as a minimum of delivery capability.

Operating Systems Compatibility

Operating Systems supported vary between providers. Due to the high frequency of change in provider delivery models, this list should be viewed as a minimum of delivery capability.

 **Q1 2011**

SOLUTION TYPE	PROVIDER TYPE | CUSTOMER CONTROL	

Serving Information
www.3par.com

PROVIDER SUMMARY

3PAR Inc. is a manufacturer of systems and software for data storage and information management headquartered in Fremont, California, USA. It is a wholly-owned subsidiary of Hewlett-Packard. 3PAR produces a range of enterprise storage products, including hardware disk arrays and storage management software. HP acquired 3PAR, the leading global provider of utility storage—a category of highly virtualized and dynamically tiered storage arrays built for public and private cloud computing and the storage building block of utility computing.

Year Founded	1999		Moody's Rating	A2		Pricing	N

IaaS Sub-Classification

White Label	
Traditional	
Servers	
Storage	✔
Disaster Recovery	
Backup	
Messaging	
Content Delivery	
Desktop-as-a-Service	
Hosting	
IT-as-a-Service	
Management Software	

Direct & Channel

Advanced Systems Group, Cintra, Quail Technologies, Veristor, Databasement, Computacenter

Clients

Terremark
Star
Verizon
Carrenza Hosting

Summary

Termination Notification (Days)	N/A
Dedicated Account Management	N/A
Live Customer Support	Y
Guaranteed Network Availability	N/A
Root/ Administrator Access	N
Portal Support	Y

Updates

Available
Q2 2011

CloudSourcing100.com

Regulation and Compliance

N/A

Programming Languages

N/A

Operating Systems Compatibility

Windows Server 2008

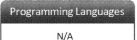

CLOUD SOURCING 100

Q1 2011

SOLUTION TYPE

PROVIDER TYPE | CUSTOMER CONTROL

www.3tera.com

PROVIDER SUMMARY

3tera, Inc., is a developer of system software for utility computing and cloud computing. It is headquartered in Aliso Viejo, California. 3tera is among the pioneers in the cloud computing space, with its AppLogic system. 3tera's flagship product, AppLogic, is used by numerous service providers as the foundation for their cloud computing offerings. AppLogic is a turnkey system that converts arrays of servers into virtualized resource pools that users can subscribe to in order to power their applications.

Year Founded	2004

Moody's Rating	Baa2

Pricing	N

IaaS Sub-Classification

White Label	
Traditional	
Servers	
Storage	
Disaster Recovery	
Backup	
Messaging	
Content Delivery	
Desktop-as-a-Service	
Hosting	
IT-as-a-Service	
Management Software	✔

Direct & Channel

Birdhosting, Carinet, ENKI, Layeredtech, SkyGone, Agathongroup, Contegix, Kualo, RightServers

Clients

Jewelry.com
DonorsChoose.org
MySatori
SilkFair
Voodoo

Summary

Termination Notification (Days)	N/A
Dedicated Account Management	N/A
Live Customer Support	Y
Guaranteed Network Availability	N/A
Root/ Administrator Access	Y
Portal Support	Y

Updates

Available
Q2 2011

CloudSourcing100.com

Regulation and Compliance

N/A

Programming Languages

N/A

Operating Systems Compatibility

FreeBSD, Linux Operating Systems, Open Solaris, Windows Server 2003, Windows Server 2008

CLOUD SOURCING 100 Q1 2011

SOLUTION TYPE

PROVIDER TYPE | CUSTOMER CONTROL

A Small Orange
www.asmallorange.com

PROVIDER SUMMARY

A Small Orange is an Infrastructure-as-a-Service provider. A Small Orange offerings include a variety of different web hosting solutions. From business web hosting to shared web hosting, from VPS (virtual private server) hosting to managed dedicated server hosting, as well as more advanced web hosting solutions. Their business hosting plans are designed specifically for high-traffic websites and online stores. They place fewer accounts on each server, include an SSL certificate and IP address with each account, and scan each server for PCI compliance.

Year Founded	2003

Moody's Rating	N/A

Pricing	Y

IaaS Sub-Classification	
White Label	
Traditional	
Servers	✔
Storage	
Disaster Recovery	
Backup	
Messaging	
Content Delivery	
Desktop-as-a-Service	
Hosting	✔
IT-as-a-Service	
Management Software	

Direct & Channel
None Listed

Clients
None Listed

Summary	
Termination Notification (Days)	N/A
Dedicated Account Management	N/A
Live Customer Support	Y
Guaranteed Network Availability	Y
Root/ Administrator Access	N
Portal Support	Y

Updates
Available Q2 2011
CloudSourcing100.com

Regulation and Compliance
N/A

Operating Systems Compatibility
N/A

Programming Languages
N/A

ALSBRIDGE

CLOUD SOURCING 100

Q1 2011

SOLUTION TYPE	PROVIDER TYPE | CUSTOMER CONTROL

Accenture
www.accenture.com

PROVIDER SUMMARY

Accenture is a global management consulting, technology services and outsourcing company. Accenture has developed the Accenture Cloud Application Factory, an industrialized delivery approach that makes it easier to either migrate existing applications or custom build new ones on a cloud platform. Accenture assists organizations with analyzing their portfolios to identify the applications that are best suited to the cloud. They also assist them with creating a plan for transforming operating models and the IT organization to become more cloud-friendly.

Year Founded	1989

Moody's Rating	A1

Pricing	N

IaaS Sub-Classification

White Label	
Traditional	
Servers	
Storage	
Disaster Recovery	
Backup	
Messaging	
Content Delivery	
Desktop-as-a-Service	
Hosting	
IT-as-a-Service	
Management Software	

Direct & Channel

None Listed

Clients

None Listed

Summary

Termination Notification (Days)	N/A
Dedicated Account Management	Y
Live Customer Support	Y
Guaranteed Network Availability	N/A
Root/ Administrator Access	N/A
Portal Support	N/A

Updates

Available
Q2 2011

CloudSourcing100.com

Regulation and Compliance

N/A

Operating Systems Compatibility

N/A

Programming Languages

N/A

CLOUDSOURCING 100   Q1 2011

SOLUTION TYPE	PROVIDER TYPE | CUSTOMER CONTROL	

www.akamai.com

PROVIDER SUMMARY

Akamai Technologies, Inc. is a company that provides a distributed computing platform for global Internet content and application delivery. Akamai's cloud optimization services help businesses improve performance, increase availability, and enhance security of applications and key web assets delivered from the Amazon EC2 infrastructure to global users. Akamai's Cloud optimization services address distance problems, network inefficiencies and congestion, thereby helping businesses realize the full benefits of their cloud strategy as a viable alternative to building out IT infrastructure.

Year Founded	1998		Moody's Rating	N/A		Pricing	N

IaaS Sub-Classification

White Label	
Traditional	
Servers	
Storage	
Disaster Recovery	
Backup	
Messaging	
Content Delivery	✓
Desktop-as-a-Service	
Hosting	
IT-as-a-Service	
Management Software	

Direct & Channel

Adobe, BT Broadcast Services, Digital River, EDS, Entriq, IBM Global Services, Kit Digital, Telefonica and many more.

Clients

NBA
Adobe
Audi
EMC
MTV

Summary

Termination Notification (Days)	N/A
Dedicated Account Management	N/A
Live Customer Support	Y
Guaranteed Network Availability	N/A
Root/ Administrator Access	N
Portal Support	N/A

Updates

Available
Q2 2011

CloudSourcing100.com

Regulation and Compliance

N/A

Operating Systems Compatibility

N/A

Programming Languages

N/A

 **Q1 2011**

SOLUTION TYPE	PROVIDER TYPE	CUSTOMER CONTROL

www.appzero.com

PROVIDER SUMMARY

AppZero™ is software used to create, control, and maintain Virtual Application Appliances (VAA). Rather than taking the approach that is typically proposed by the suppliers of virtual machine technology, AppZero's approach is to encapsulate applications rather than complete desktop or server environments. Application virtualization software suppliers, such as AppZero, suggest that this approach means that the images projected into the Cloud would be smaller, easier to manage and can be made compatible with a larger number of suppliers of cloud computing infrastructure.

Year Founded	N/A	Moody's Rating	N/A	Pricing	N

IaaS Sub-Classification

White Label	
Traditional	
Servers	
Storage	
Disaster Recovery	
Backup	
Messaging	
Content Delivery	
Desktop-as-a-Service	
Hosting	
IT-as-a-Service	
Management Software	✔

Direct & Channel

Amazon Web Services, GoGrid, Op-Source, Hewlett-Packard, Microsoft, Sun, Novell, VMWare, RedHat

Clients

None Listed

Summary

Termination Notification (Days)	N/A
Dedicated Account Management	N/A
Live Customer Support	Y
Guaranteed Network Availability	N/A
Root/ Administrator Access	N
Portal Support	N

Updates

Available
Q2 2011

CloudSourcing100.com

Regulation and Compliance

N/A

Programming Languages

N/A

Operating Systems Compatibility

Windows 2003, Windows 2008R1, Windows 2008, Appliance support (Sparc Solaris 2.6, 7, 8,9 and 10) onto Sparc Solaris 9 and 10, Red Hat Enterprise and SUSE Linux Enterprise from Novell

ALSBRIDGE

 CING 100   **Q1 2011**

SOLUTION TYPE	PROVIDER TYPE | CUSTOMER CONTROL	

Arjuna
www.arjuna.com

PROVIDER SUMMARY

Arjuna Technologies Ltd. is a technology consultancy and product development business which specializes in Cloud Computing. They deliver scalable, fault tolerant software products to global organisations. The company offers Arjuna Agility, a federated cloud computing platform that allows organizations to deploy cloud without disrupting existing infrastructure and applications to improve business agility, as well as supports Web services using SOAP/HTTP. Arjuna also provides middleware technology for distributed systems; & consultancy, technology, & the integration support services to platform vendors and the enterprise class companies.

Year Founded	2002		Moody's Rating	N/A		Pricing	N

IaaS Sub-Classification

White Label	
Traditional	
Servers	
Storage	
Disaster Recovery	
Backup	
Messaging	
Content Delivery	
Desktop-as-a-Service	
Hosting	
IT-as-a-Service	
Management Software	✓

Direct & Channel

None Listed

Clients

None Listed

Summary

Termination Notification (Days)	N/A
Dedicated Account Management	N/A
Live Customer Support	N/A
Guaranteed Network Availability	N/A
Root/ Administrator Access	N
Portal Support	N

Updates

Available
Q2 2011

CloudSourcing100.com

Regulation and Compliance

N/A

Operating Systems Compatibility

N/A

Programming Languages

N/A

ALSBRIDGE

CLOUD SOURCING 100   Q1 2011

SOLUTION TYPE	PROVIDER TYPE | CUSTOMER CONTROL	

Asankya
www.asankya.com

PROVIDER SUMMARY

Asankya is the cloud acceleration company for high speed delivery of Internet-based applications. The company has built a Cloud Acceleration Network (CAN) service currently in final beta testing with leading SaaS companies, cloud storage providers, internal enterprise cloud users and key government entities. Asankya's RAPIDnet enables performance improvement for Internet-based applications, including cloud storage, real-time collaboration, SaaS, virtual desktop infrastructure and interactive video, providing the performance of private networks over the Internet.

Year Founded	2004

Moody's Rating	N/A

Pricing	N

IaaS Sub-Classification	
White Label	
Traditional	
Servers	
Storage	
Disaster Recovery	
Backup	
Messaging	
Content Delivery	✔
Desktop-as-a-Service	
Hosting	
IT-as-a-Service	
Management Software	

Direct & Channel
None Listed

Clients
None Listed

Summary	
Termination Notification (Days)	N/A
Dedicated Account Management	N/A
Live Customer Support	N/A
Guaranteed Network Availability	N/A
Root/ Administrator Access	N
Portal Support	N/A

Updates
Available Q2 2011
CloudSourcing100.com

Regulation and Compliance
N/A

Operating Systems Compatibility
N/A

Programming Languages
N/A

ALSBRIDGE

CLOUD SOURCING 100

Q1 2011

SOLUTION TYPE

PROVIDER TYPE | CUSTOMER CONTROL

AT&T
www.synaptic.att.com

PROVIDER SUMMARY

AT&T offers Intelligent Content Distribution services. AT&T's Synaptic Compute as a Service, is a cloud-based, pay-as-you-go, on-demand service that lets you quickly obtain virtual servers for your business and dial the service up or down in minutes. Their Synaptic Storage as a Service is a web services-based storage solution that easily scales up and down to any size you need and allows you to pay only for the storage you use.

Year Founded	1983

Moody's Rating	A2

Pricing	Y

IaaS Sub-Classification

White Label	
Traditional	
Servers	✓
Storage	✓
Disaster Recovery	
Backup	
Messaging	
Content Delivery	
Desktop-as-a-Service	
Hosting	
IT-as-a-Service	
Management Software	

Direct & Channel

Cirtas, Commvault, Dolphin, Gladinet, Nasuni, Riverbed, Seven10, Storage-Point, Storsimple, TwinStrata

Clients

None Specific to Cloud

Summary

Termination Notification (Days)	N/A
Dedicated Account Management	N/A
Live Customer Support	Y
Guaranteed Network Availability	Y
Root/ Administrator Access	N
Portal Support	Y

Updates

Available Q2 2011

CloudSourcing100.com

Regulation and Compliance

N/A

Programming Languages

N/A

Operating Systems Compatibility

Windows 2008 (64 bit) and Red Hat Linux Enterprise (64 bit)

 CLOUD SOURCING 100 **Q1 2011**

SOLUTION TYPE	PROVIDER TYPE | CUSTOMER CONTROL	
I P S B	**B**	**H**

www.aws.amazon.com

PROVIDER SUMMARY

The Amazon Web Services (AWS) are a collection of remote computing services that together make up a cloud computing platform, offered over the Internet by Amazon.com. The most central and well-known of these services are Amazon EC2 and Amazon S3. Amazon Web Services provide online services for other web sites or client-side applications. Most of these services are not exposed directly to end users, but instead offer functionality that other developers can use. Amazon Web Services' offerings are accessed over HTTP, using REST and SOAP protocols. All are billed on usage, with the exact form of usage varying from service to service.

Year Founded	2002

Moody's Rating	Baa3

Pricing	Y

IaaS Sub-Classification

White Label	✔
Traditional	
Servers	✔
Storage	✔
Disaster Recovery	
Backup	
Messaging	✔
Content Delivery	✔
Desktop-as-a-Service	
Hosting	
IT-as-a-Service	
Management Software	

Direct & Channel

8kMiles, AppZero, Citrix, CloudSwitch, Cloudera, F5 Networks, HP, IBM, Layer 7, Novell

Summary

Termination Notification (Days)	15
Dedicated Account Management	N
Live Customer Support	Y
Guaranteed Network Availability	Y
Root/ Administrator Access	Y
Portal Support	Y

Clients

6 Waves
Envoy Media Group
RedBus.in
Etsy
Virgin Atlantic

Updates

Available
Q2 2011

CloudSourcing100.com

Regulation and Compliance

ISO 27001, SAS70 TYPE II, PCI/DSS LEVEL1, HIPPA

Programming Languages

Java

Operating Systems Compatibility

RedHat Linux, Windows Server, openSuSE Linux, Fedora, Debian, OpenSolaris, Cent OS, Gentoo Linux, and Oracle Linux

 ALSBRIDGE

CLOUD SOURCING 100 — Q1 2011

SOLUTION TYPE

I P S B

PROVIDER TYPE | CUSTOMER CONTROL

B L

Boomi
www.boomi.com

PROVIDER SUMMARY

Dell Boomi is a provider of on-demand integration technology and the creator of AtomSphere®. AtomSphere connects providers and customers of SaaS, cloud and on-premise applications via a SaaS integration platform that does not require software or appliances. ISVs and businesses alike benefit by connecting to the network of SaaS, PaaS, on-premise and cloud computing environments in a seamless and self-service model.

Year Founded	2000

Moody's Rating	A2

Pricing	N

IaaS Sub-Classification

White Label	
Traditional	
Servers	
Storage	
Disaster Recovery	
Backup	
Messaging	
Content Delivery	
Desktop-as-a-Service	
Hosting	
IT-as-a-Service	
Management Software	

Direct & Channel

AWS, Appnexus, Intuit, OpSource, Varien

Clients

SalesForce.com
NetSuite
RightNow
Taleo
NASDAQ

Summary

Termination Notification (Days)	N/A
Dedicated Account Management	N/A
Live Customer Support	N
Guaranteed Network Availability	N/A
Root/ Administrator Access	N
Portal Support	N

Updates

Available
Q2 2011

CloudSourcing100.com

Regulation and Compliance

N/A

Operating Systems Compatibility

Cent OS, Windows Server 2003, Windows Server 2008

Programming Languages

AS400, J D Edwards, Lotus Notes & Domino & more

ALSBRIDGE

CLOUD SOURCING 100 Q1 2011

SOLUTION TYPE	PROVIDER TYPE | CUSTOMER CONTROL	
		 www.cachefly.com

PROVIDER SUMMARY

CacheFly is a content delivery network provider based in Chicago, IL. CacheNetworks offers the CacheFly CDN, which uses BGP AnyCast to find the server with lowest latency to the user. Cache-Fly enables companies to rapidly expand without adding infrastructure while simultaneously increasing performance and reducing costs. The CacheFly content delivery network gives customers on-demand capacity, unsurpassed performance and a predictable cost model.

Year Founded	2002		Moody's Rating	N/A		Pricing	N

IaaS Sub-Classification

White Label	
Traditional	
Servers	
Storage	
Disaster Recovery	
Backup	
Messaging	
Content Delivery	✔
Desktop-as-a-Service	
Hosting	
IT-as-a-Service	
Management Software	

Direct & Channel

None Listed

Clients

CrowdScience
Wizzard
Quark
Budweiser
Honda

Summary

Termination Notification (Days)	N/A
Dedicated Account Management	N/A
Live Customer Support	Y
Guaranteed Network Availability	Y
Root/ Administrator Access	N
Portal Support	N/A

Updates

Available
Q2 2011

CloudSourcing100.com

Regulation and Compliance

SAS70 Type II

Operating Systems Compatibility

N/A

Programming Languages

N/A

CLOUD SOURCING 100

Q1 2011

SOLUTION TYPE	PROVIDER TYPE	CUSTOMER CONTROL

Capgemini
www.capgemini.com

PROVIDER SUMMARY

Capgemini is one of the world's largest management consulting, outsourcing and professional services company. Since the emergence of cloud as a legitimate IT platform for enterprises, Capgemini has been at the forefront of bringing cloud computing to the enterprise. They work with partners such as Amazon Web Services, Google, IBM and Microsoft to offer a scalable, dependable cloud-based compute and storage services to end-users.

Year Founded	1967		Moody's Rating	N/A		Pricing	N

IaaS Sub-Classification

White Label	
Traditional	✔
Servers	✔
Storage	
Disaster Recovery	
Backup	
Messaging	
Content Delivery	
Desktop-as-a-Service	
Hosting	
IT-as-a-Service	
Management Software	

Direct & Channel

Amazon Web Services

Clients

None Specific to Cloud

Summary

Termination Notification (Days)	NG
Dedicated Account Management	Y
Live Customer Support	Y
Guaranteed Network Availability	Y
Root/ Administrator Access	Y
Portal Support	Y

Updates

Available Q2 2011

CloudSourcing100.com

Regulation and Compliance

Multiple

Operating Systems Compatibility

N/A

Programming Languages

None Listed

ALSBRIDGE

© Alsbridge Inc, 2011

CLOUD SOURCING 100   Q1 2011

SOLUTION TYPE	PROVIDER TYPE | CUSTOMER CONTROL	
		 www.carpathia.com

PROVIDER SUMMARY

Carpathia's InstantOn™ brings together best in breed technology including Citrix XenServer, Citrix Netscaler, 3par, Juniper, Dell, RSA and Highwinds to deliver cloud solutions that meet the most demanding enterprise and federal workloads. Their public cloud is offered with full 24x7 support from Carpathia's operations team and onboarding of your applications, paired with the ability to elect virtual machines to be monitored by the Carpathia operations center.

Year Founded	2003

Moody's Rating	N/A

Pricing	N

IaaS Sub-Classification

White Label	
Traditional	
Servers	✓
Storage	
Disaster Recovery	
Backup	
Messaging	
Content Delivery	
Desktop-as-a-Service	
Hosting	
IT-as-a-Service	
Management Software	

Regulation and Compliance

SAS70 Type II, Safeharbor

Programming Languages

N/A

Direct & Channel

Citrix, ParaScale, HighWinds, Juniper Networks, RSA

Summary

Termination Notification (Days)	15
Dedicated Account Management	N/A
Live Customer Support	Y
Guaranteed Network Availability	Y
Root/ Administrator Access	Y
Portal Support	N/A

Operating Systems Compatibility

N/A

Clients

Northrop Grumann
Ceridian
Saba
SugarSynch
Ventraq

Updates

Available Q2 2011

CloudSourcing100.com

ALSBRIDGE

 Q1 2011

SOLUTION TYPE

PROVIDER TYPE | CUSTOMER CONTROL

www.us.cdnetworks.com

PROVIDER SUMMARY

CDNetworks founded in 2000, is a full service content delivery network (CDN), with increasing business in the United States. CDNetworks, a full-service content delivery network (CDN), provides technology, network infrastructure, and customer services for the delivery of Internet content and applications.It provides technology for video streaming, large-volume-files downloads, and image caching.

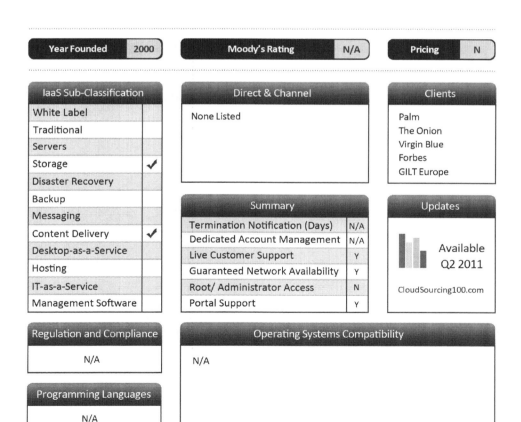

Year Founded	2000

Moody's Rating	N/A

Pricing	N

IaaS Sub-Classification

White Label	
Traditional	
Servers	
Storage	✓
Disaster Recovery	
Backup	
Messaging	
Content Delivery	✓
Desktop-as-a-Service	
Hosting	
IT-as-a-Service	
Management Software	

Direct & Channel

None Listed

Summary

Termination Notification (Days)	N/A
Dedicated Account Management	N/A
Live Customer Support	Y
Guaranteed Network Availability	Y
Root/ Administrator Access	N
Portal Support	Y

Clients

Palm
The Onion
Virgin Blue
Forbes
GILT Europe

Updates

Available
Q2 2011

CloudSourcing100.com

Regulation and Compliance

N/A

Operating Systems Compatibility

N/A

Programming Languages

N/A

ALSBRIDGE

CLOUD SOURCING 100

Q1 2011

SOLUTION TYPE	PROVIDER TYPE | CUSTOMER CONTROL	
I P S B	F H	 www.citrix.com

PROVIDER SUMMARY

Citrix Systems, Inc. is a multinational corporation, that provides server and desktop virtualiza-tion, networking, software-as-a-service (SaaS) and cloud computing technologies including Xen open source products. Citrix currently services around 230,000 organizations worldwide. Citrix' key product families are Citrix Delivery Center, Citrix Cloud Center (C3) and Citrix Online Services product families.The Citrix C3 solution integrates "cloud proven" virtualization and networking products that power many of today's largest Internet and web service providers.

Year Founded	1989

Moody's Rating	N/A

Pricing	N

IaaS Sub-Classification	
White Label	
Traditional	
Servers	
Storage	
Disaster Recovery	
Backup	
Messaging	
Content Delivery	
Desktop-as-a-Service	
Hosting	
IT-as-a-Service	
Management Software	✔

Direct & Channel
Mulitple Partner Listings

Clients
Multiple Global Clients

Summary	
Termination Notification (Days)	N/A
Dedicated Account Management	N/A
Live Customer Support	Y
Guaranteed Network Availability	N/A
Root/ Administrator Access	Y
Portal Support	Y

Updates
Available Q2 2011
CloudSourcing100.com

Regulation and Compliance
N/A

Programming Languages
N/A

Operating Systems Compatibility
Windows Server 2003, Windows Server 2008

ALSBRIDGE

 **Q1 2011**

SOLUTION TYPE	PROVIDER TYPE | CUSTOMER CONTROL	

www.cloudshare.com

PROVIDER SUMMARY

CloudShare is a cloud computing provider which enables users to create, replicate and share fully-functional IT environments in the Cloud. CloudShare combines aspects of virtualization, cloud computing and web conferencing to offer a software as a service (Saas) solution for delivering IT to colleagues, clients, customers and partners. Similar to virtual lab automation, Cloud-Share makes full-featured virtual enterprise environments available online and on-demand.

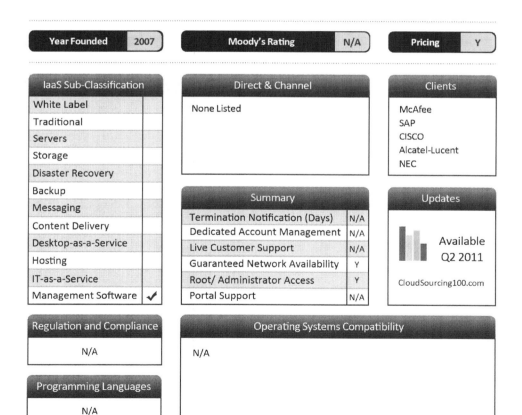

| Year Founded | 2007 | | Moody's Rating | N/A | | Pricing | Y |

IaaS Sub-Classification

White Label	
Traditional	
Servers	
Storage	
Disaster Recovery	
Backup	
Messaging	
Content Delivery	
Desktop-as-a-Service	
Hosting	
IT-as-a-Service	
Management Software	✓

Direct & Channel

None Listed

Clients

McAfee
SAP
CISCO
Alcatel-Lucent
NEC

Summary

Termination Notification (Days)	N/A
Dedicated Account Management	N/A
Live Customer Support	N/A
Guaranteed Network Availability	Y
Root/ Administrator Access	Y
Portal Support	N/A

Updates

Available
Q2 2011

CloudSourcing100.com

Regulation and Compliance

N/A

Operating Systems Compatibility

N/A

Programming Languages

N/A

CLOUD SOURCING 100

Q1 2011

SOLUTION TYPE	PROVIDER TYPE	CUSTOMER CONTROL

www.cloudsigma.com

PROVIDER SUMMARY

CloudSigma is an Infrastructure-as-a-Service (IaaS) company based in Zurich, Switzerland. CloudSigma provides cloud servers on a utility computing basis adopting a pure Infrastructure-as-a-Service approach with open networking and an open software later. Their cloud servers are managed via its web based provisioning portal or via API. CloudSigma's stated aim is to provide a pure IaaS product initially targeting the continental European market.

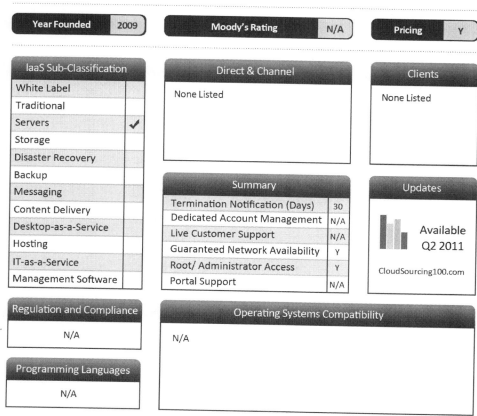

Year Founded	2009		Moody's Rating	N/A		Pricing	Y

IaaS Sub-Classification

White Label	
Traditional	
Servers	✓
Storage	
Disaster Recovery	
Backup	
Messaging	
Content Delivery	
Desktop-as-a-Service	
Hosting	
IT-as-a-Service	
Management Software	

Direct & Channel

None Listed

Clients

None Listed

Summary

Termination Notification (Days)	30
Dedicated Account Management	N/A
Live Customer Support	N/A
Guaranteed Network Availability	Y
Root/ Administrator Access	Y
Portal Support	N/A

Updates

Available Q2 2011

CloudSourcing100.com

Regulation and Compliance

N/A

Operating Systems Compatibility

N/A

Programming Languages

N/A

ALSBRIDGE

 CING 100 **Q1 2011**

SOLUTION TYPE	PROVIDER TYPE | CUSTOMER CONTROL	

www.cloudswitch.com

PROVIDER SUMMARY

CloudSwitch delivers the enterprise gateway to the cloud. CloudSwitch's innovative software appliance enables enterprises to run their applications in the cloud computing environment-securely, simply and without changes. CloudSwitch protects enterprises from the complexity, risks and potential lock-in of cloud computing, freeing them to leverage the cloud's advantages in cost and business agility. Backed by Matrix Partners, Atlas Ventures and Commonwealth Capital Ventures, CloudSwitch is based in Burlington, MA.

Year Founded	2008		Moody's Rating	N/A		Pricing	N

IaaS Sub-Classification	
White Label	
Traditional	
Servers	
Storage	
Disaster Recovery	
Backup	
Messaging	
Content Delivery	
Desktop-as-a-Service	
Hosting	
IT-as-a-Service	
Management Software	✓

Direct & Channel
None Listed

Clients
AWS
Microsoft
Terremark
VMWARE

Summary	
Termination Notification (Days)	N/A
Dedicated Account Management	N/A
Live Customer Support	Y
Guaranteed Network Availability	Y
Root/ Administrator Access	Y
Portal Support	N/A

Updates
Available Q2 2011
CloudSourcing100.com

Regulation and Compliance
N/A

Operating Systems Compatibility
RedHat/CentOS/Oracle Enterprise Linux 5.x (x86 & x64), Windows Server 2003/2008 (x86 & x64)

Programming Languages
Java

CLOUD SOURCING 100 Q1 2011

SOLUTION TYPE	PROVIDER TYPE	CUSTOMER CONTROL

Cloudworks
www.cloudworks.com

PROVIDER SUMMARY

Appirio CloudWorks enables enterprises to unleash information from SaaS silos and deliver it through the tools business people use to work. CloudWorks is cloud broker technology, delivered as a service, that unifies identity, security, data, context, and business object definitions across leading cloud applications such as Salesforce, Google Apps, Workday and their ecosystems. This makes cross-cloud solutions possible and makes it significantly easier to manage data, users and processes across SaaS applications.

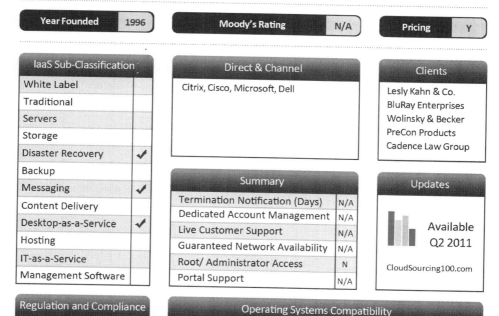

Year Founded	1996

Moody's Rating	N/A

Pricing	Y

IaaS Sub-Classification

White Label	
Traditional	
Servers	
Storage	
Disaster Recovery	✓
Backup	
Messaging	✓
Content Delivery	
Desktop-as-a-Service	✓
Hosting	
IT-as-a-Service	
Management Software	

Direct & Channel

Citrix, Cisco, Microsoft, Dell

Clients

Lesly Kahn & Co.
BluRay Enterprises
Wolinsky & Becker
PreCon Products
Cadence Law Group

Summary

Termination Notification (Days)	N/A
Dedicated Account Management	N/A
Live Customer Support	N/A
Guaranteed Network Availability	N/A
Root/ Administrator Access	N
Portal Support	N/A

Updates

Available
Q2 2011

CloudSourcing100.com

Regulation and Compliance

N/A

Operating Systems Compatibility

Windows Server 2003

Programming Languages

N/A

CLOUD SOURCING 100   Q1 2011

SOLUTION TYPE	PROVIDER TYPE | CUSTOMER CONTROL	

Cognizant
www.cognizant.com

PROVIDER SUMMARY

Cognizant provides a range of information technology, consulting and Business Processing Outsourcing (BPO) services. Cognizant was one of the first IT services companies to organize around key industry verticals as well as technology horizontals. Cognizant's cloud based services include Business Process as a Service (BPaaS), Cloud Strategy, Cloud Advisory Services,and Cloud Migration Management.

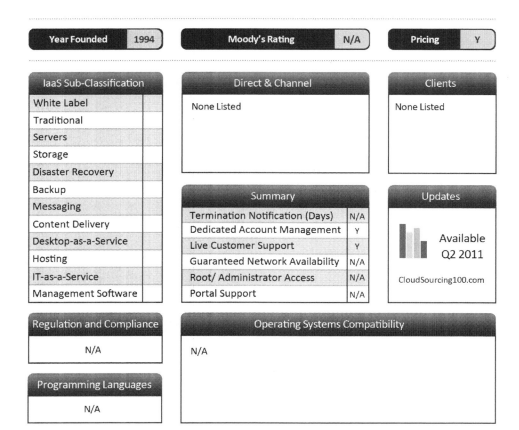

Year Founded	1994

Moody's Rating	N/A

Pricing	Y

IaaS Sub-Classification
White Label	
Traditional	
Servers	
Storage	
Disaster Recovery	
Backup	
Messaging	
Content Delivery	
Desktop-as-a-Service	
Hosting	
IT-as-a-Service	
Management Software	

Direct & Channel
None Listed

Clients
None Listed

Summary
Termination Notification (Days)	N/A
Dedicated Account Management	Y
Live Customer Support	Y
Guaranteed Network Availability	N/A
Root/ Administrator Access	N/A
Portal Support	N/A

Updates
Available
Q2 2011

CloudSourcing100.com

Regulation and Compliance
N/A

Operating Systems Compatibility
N/A

Programming Languages
N/A

CLOUD SOURCING 100   Q1 2011

SOLUTION TYPE	PROVIDER TYPE | CUSTOMER CONTROL	
		 www.cohesiveft.com

PROVIDER SUMMARY

CohesiveFT's Cloud Container process is a methodology to take a client's aspirational topology, i.e., the relevant aspects of their existing physical infrastructure, and create a Cloud Container Solution to enable the rapid on-demand deployment of the cluster topology in cloud environments. Their VPN-Cubed Cloud Only solution provides users flexibility with control in public cloud environments. They currently offer self-service Editions for use in Amazon EC2, GoGrid, Terremark, and Flexiant's Flexiscle.

Year Founded	2006	Moody's Rating	N/A	Pricing	N

IaaS Sub-Classification

White Label	
Traditional	
Servers	
Storage	
Disaster Recovery	
Backup	
Messaging	
Content Delivery	
Desktop-as-a-Service	
Hosting	
IT-as-a-Service	
Management Software	✔

Direct & Channel

Amazon Web Services, Elastic Hosts, Eucalyptus Systems, Flexiscale, Gigaspaces, GoGrid, IBM, Rackspace, Skytap, Citrix, Paralells, VMWARE

Clients

None Listed

Summary

Termination Notification (Days)	N/A
Dedicated Account Management	N/A
Live Customer Support	N
Guaranteed Network Availability	N
Root/ Administrator Access	N
Portal Support	Y

Updates

Available Q2 2011

CloudSourcing100.com

Regulation and Compliance

N/A

Operating Systems Compatibility

N/A

Programming Languages

N/A

ALSBRIDGE

CLOUD  SOURCING 100   Q1 2011

SOLUTION TYPE	PROVIDER TYPE	CUSTOMER CONTROL

smarter / faster / further
www.colt.net

PROVIDER SUMMARY

Colt Managed Services is an Infrastructure-as-Service (IaaS) provider offering services that are built on infrastructure covering 25,000km of network and 19 data centres across Europe. This scale, coupled with ownership of the assets, means they take responsibility, end-to-end, for the security, availability and continuity of your IT processes and computing capability. Furthermore, they underwrite the service levels they promise with financial guarantees.

Year Founded	1992

Moody's Rating	N/A

Pricing	N

IaaS Sub-Classification

White Label	
Traditional	
Servers	✔
Storage	✔
Disaster Recovery	
Backup	✔
Messaging	✔
Content Delivery	
Desktop-as-a-Service	
Hosting	
IT-as-a-Service	
Management Software	

Direct & Channel

None Listed

Summary

Termination Notification (Days)	N/A
Dedicated Account Management	Y
Live Customer Support	Y
Guaranteed Network Availability	Y
Root/ Administrator Access	N
Portal Support	N/A

Clients

Sdu Uitgevers
Direct Line
Versicherung
IDG
Digitick

Updates

Available
Q2 2011

CloudSourcing100.com

Regulation and Compliance

N/A

Operating Systems Compatibility

N/A

Programming Languages

N/A

ALSBRIDGE

CLOUD SOURCING 100

Q1 2011

SOLUTION TYPE	PROVIDER TYPE | CUSTOMER CONTROL

CompuCom.
The Leading IT Outsourcing Specialist
www.compucom.com

PROVIDER SUMMARY

CompuCom Systems, Inc. Corp. is a US-based Information Technology Outsourcing company, specializing in Infrastructure Management, Applications & Software Development Services, Systems integration, and IT Workspace Management. CompuCom offers IT infrastructure management services, application development, systems integration, consulting and professional services, as well as the procurement & management of software & hardware. CompuCom's flagship solution & framework helps organizations reduce operating & capital expense, drive alignment of IT services to business plans, & improve the value contribution of innovative technologies & processes.

Year Founded	2006	Moody's Rating	N/A	Pricing	N

IaaS Sub-Classification

White Label	
Traditional	✔
Servers	✔
Storage	✔
Disaster Recovery	
Backup	✔
Messaging	✔
Content Delivery	
Desktop-as-a-Service	✔
Hosting	✔
IT-as-a-Service	✔
Management Software	

Direct & Channel

None Listed

Summary

Termination Notification (Days)	30
Dedicated Account Management	Y
Live Customer Support	Y
Guaranteed Network Availability	Y
Root/ Administrator Access	Y
Portal Support	Y

Clients

Oracle
Hewlett-Packard
Acresso
Ellie Mae
SAVVIS

Updates

Available
Q2 2011

CloudSourcing100.com

Regulation and Compliance

SAS70 Type II

Programming Languages

SQL, MySQL, Sharepoint
Joomla, Selenium & more

Operating Systems Compatibility

Windows Clients (Vista, XP, etc.), Windows Servers (2000, 2003, 2008, 32 and 64 bit etc.), Linux (Red Hat Enterprise, Ubuntu, CentOS, Debian, etc.)

ALSBRIDGE

 C I N G 100 **Q1 2011**

SOLUTION TYPE	PROVIDER TYPE	CUSTOMER CONTROL	

CONTEGIX
www.contegix.com

PROVIDER SUMMARY

Contegix provides high-level managed hosting solutions for enterprise applications and infrastructure. Contegix Cloud services are enterprise-ready cloud services that address issues of consistent performance, migration, compliance, management, and SLAs that allow even regulated industries to use the power and elasticity of modern cloud computing. Powered by VMware ESX hypervisors, Contegix Cloud Services enables you to migrate physical hosts to virtual hosts, use approved appliances, and create your own templates through a well established and understood set of tools.

Year Founded	2004		Moody's Rating	N/A		Pricing	Y

IaaS Sub-Classification	
White Label	
Traditional	
Servers	✔
Storage	
Disaster Recovery	
Backup	
Messaging	
Content Delivery	
Desktop-as-a-Service	
Hosting	✔
IT-as-a-Service	
Management Software	

Direct & Channel

Relevance, 3Tera, Six Apart, Sonatype, VMWare

Summary

Termination Notification (Days)	N/A
Dedicated Account Management	N/A
Live Customer Support	Y
Guaranteed Network Availability	Y
Root/ Administrator Access	Y
Portal Support	N/A

Clients

Atlassian
AppFuse
Ticketfly
TalentQuest
Hotwax Media

Updates

Available
Q2 2011

CloudSourcing100.com

Regulation and Compliance

SAS70 Type II, HIPPA, Safe Harbor and more

Operating Systems Compatibility

Linux, Mac OS X

Programming Languages

N/A

ALSBRIDGE

 **Q1 2011**

SOLUTION TYPE	PROVIDER TYPE | CUSTOMER CONTROL	

www.cordys.com

PROVIDER SUMMARY

The Cordys Business Operations Platform is web-based and fully SaaS enabled, with no client implementation requirements other than a web browser. Cordys Cloud Provisioning complements the Business Operations Platform with automated provisioning & metering of applications for the Cloud. This is a single platform, which allows organizations to design, execute, monitor, change and optimize critical business processes and operations. This platform is also a software stack that consists of Enterprise Service Bus (ESB), Business Process Management, Workflow, Master Data Management, Business Activity Monitoring, and enterprise mashup framework.

Year Founded	2001		Moody's Rating	N/A		Pricing	N

IaaS Sub-Classification

White Label
Traditional
Servers
Storage
Disaster Recovery
Backup
Messaging
Content Delivery
Desktop-as-a-Service
Hosting
IT-as-a-Service
Management Software

Direct & Channel

Accenture, Atos Origin, Capgemini, CSC, Wipro

Clients

Comcast
NYSE
Savvis
Essar Group
Skyworth

Summary

Termination Notification (Days)	N/A
Dedicated Account Management	N/A
Live Customer Support	N/A
Guaranteed Network Availability	N/A
Root/ Administrator Access	N
Portal Support	Y

Updates

Available
Q2 2011

CloudSourcing100.com

Regulation and Compliance

N/A

Operating Systems Compatibility

Ubuntu Linux

Programming Languages

C, C++, Java, SQL

ALSBRIDGE

CLOUD SOURCING 100

Q1 2011

SOLUTION TYPE	PROVIDER TYPE | CUSTOMER CONTROL	

CSC
www.csc.com/cloud

PROVIDER SUMMARY

CSC BizCloud™ is a private, on-premises cloud built by CSC, ready for workload deployment in just 10 weeks and billed as a service from a standard rate card. BizCloud combines the privacy, security and control of a private cloud with the commercial model, elasticity and convenience of a public cloud.

Year Founded	1959		Moody's Rating	Baa3		Pricing	N

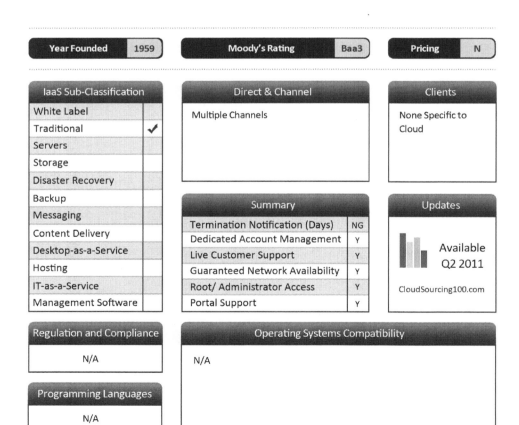

IaaS Sub-Classification

White Label	
Traditional	✔
Servers	
Storage	
Disaster Recovery	
Backup	
Messaging	
Content Delivery	
Desktop-as-a-Service	
Hosting	
IT-as-a-Service	
Management Software	

Direct & Channel

Multiple Channels

Clients

None Specific to Cloud

Summary

Termination Notification (Days)	NG
Dedicated Account Management	Y
Live Customer Support	Y
Guaranteed Network Availability	Y
Root/ Administrator Access	Y
Portal Support	Y

Updates

Available
Q2 2011

CloudSourcing100.com

Regulation and Compliance

N/A

Operating Systems Compatibility

N/A

Programming Languages

N/A

ALSBRIDGE

 **Q1 2011**

SOLUTION TYPE	PROVIDER TYPE | CUSTOMER CONTROL	
		Cumulux

www.cumulux.com

PROVIDER SUMMARY

Cumulux offers product and services that enable enterprises strategize, develop and operational-ize Cloud Computing applications. Cumulux has been working on the Windows Azure Platform since its Alpha days for more than 18 months. Cumulux has created several tools and templates to quickly assess the cloud fit of clients portfolio, model the economics of solutions, design patterns that accelerate cloud development and critical KPIs to monitor clients application on the cloud. Cumulux's two products Hybrid Axis and Mobile Axis provide the extensibility that allows salesforce.com customers to surface customer data in different contexts.

Year Founded	2008

Moody's Rating	N/A

Pricing	N

IaaS Sub-Classification
White Label
Traditional
Servers
Storage
Disaster Recovery
Backup
Messaging
Content Delivery
Desktop-as-a-Service
Hosting
IT-as-a-Service
Management Software

Direct & Channel
None Listed

Clients
Ford
Formotus
UNICEF
Microsoft

Summary
Termination Notification (Days)	N/A
Dedicated Account Management	N/A
Live Customer Support	N/A
Guaranteed Network Availability	N/A
Root/ Administrator Access	N
Portal Support	N/A

Updates
Available
Q2 2011

CloudSourcing100.com

Regulation and Compliance
N/A

Operating Systems Compatibility
N/A

Programming Languages
N/A

ALSBRIDGE

 **CLOUD** SOURCING 100 **Q1 2011**

SOLUTION TYPE

PROVIDER TYPE | CUSTOMER CONTROL

Dell
www.perotsystems.com

PROVIDER SUMMARY

Dell has developed a growing portfolio of efficient cloud computing solutions to achieve that flexible infrastructure. They offer consulting, cloud computing components and turnkey solutions based on pretested, preassembled, fully-supported hardware, software and services. With Dell's Cloud Integration Services, users benefit from their mature integration processes and disciplined methodologies. By combining their comprehensive sourcing strategies that employ best-of-breed offerings, obtained internally or through Dell's extensive team of alliances, they help users take full advantage of pay-per-use IT services.

Year Founded	1988

Moody's Rating	A2

Pricing	N

IaaS Sub-Classification

White Label	
Traditional	✓
Servers	✓
Storage	✓
Disaster Recovery	✓
Backup	✓
Messaging	✓
Content Delivery	
Desktop-as-a-Service	✓
Hosting	✓
IT-as-a-Service	✓
Management Software	

Direct & Channel

None Listed

Clients

None Specific to Cloud

Summary

Termination Notification (Days)	N/A
Dedicated Account Management	Y
Live Customer Support	Y
Guaranteed Network Availability	Y
Root/ Administrator Access	Y
Portal Support	N/A

Updates

Available
Q2 2011

CloudSourcing100.com

Regulation and Compliance

N/A

Operating Systems Compatibility

N/A

Programming Languages

N/A

CLOUD SOURCING 100

Q1 2011

SOLUTION TYPE	PROVIDER TYPE | CUSTOMER CONTROL

Dotblock
www.dotblock.com

PROVIDER SUMMARY

Dotblock is a privately owned, multi-million dollar corporation located in Upstate New York. Dot-Block Cloud VPS Hosting started up in 1999 and they are now serving around 50,000 websites. Their focus is on cloud computing and they offer their clients both Cloud VPS Hosting and Reseller VPS Hosting. Their plans are scalable, of course, which means that if your website grows, and with it its needs, so will your hosting. By focusing on service while maintaining affordability, the company provides service that meets the unique needs of each client and exceeds their wildest performance expectations.

Year Founded	1999		Moody's Rating	N/A		Pricing	Y

IaaS Sub-Classification

White Label	✓
Traditional	
Servers	✓
Storage	
Disaster Recovery	
Backup	
Messaging	
Content Delivery	
Desktop-as-a-Service	
Hosting	
IT-as-a-Service	
Management Software	

Direct & Channel

None Listed

Clients

None Listed

Summary

Termination Notification (Days)	N/A
Dedicated Account Management	N/A
Live Customer Support	Y
Guaranteed Network Availability	Y
Root/ Administrator Access	Y
Portal Support	Y

Updates

Available
Q2 2011

CloudSourcing100.com

Regulation and Compliance

N/A

Programming Languages

N/A

Operating Systems Compatibility

CentOS VPS Hosting, Fedora VPS Hosting, Debian VPS Hosting, Gentoo VPS Hosting, Slackware VPS Hosting, Ubuntu VPS Hosting, Windows 2008 VPS Hosting and Windows 2003 VPS Hosting.

ALSBRIDGE

 **CLOUD** SOURCING 100 **Q1 2011**

SOLUTION TYPE	PROVIDER TYPE | CUSTOMER CONTROL	

www.edgecast.com

PROVIDER SUMMARY

EdgeCast is a content delivery network (CDN) offering Flash, Windows, Silverlight and HTTP Progressive download streaming. The company uses technology entrepreneurs with years of experience for building companies in the infrastructure, web services, and application delivery spaces. They also provide website acceleration for increasing web site performance and speeding up page load times as well as advanced reporting and analytics.

Year Founded	2006		Moody's Rating	N/A		Pricing	N

IaaS Sub-Classification

White Label	
Traditional	
Servers	
Storage	
Disaster Recovery	
Backup	
Messaging	
Content Delivery	✓
Desktop-as-a-Service	
Hosting	
IT-as-a-Service	
Management Software	

Direct & Channel

None Listed

Summary

Termination Notification (Days)	N/A
Dedicated Account Management	N/A
Live Customer Support	N/A
Guaranteed Network Availability	Y
Root/ Administrator Access	N
Portal Support	N/A

Clients

ESPN
Yahoo!
LinkedIn
Kellogg's
JetBlue

Updates

Available
Q2 2011

CloudSourcing100.com

Regulation and Compliance

N/A

Programming Languages

N/A

Operating Systems Compatibility

N/A

CLOUD SOURCING 100 — Q1 2011

SOLUTION TYPE	PROVIDER TYPE	CUSTOMER CONTROL	
			ElasticHosts

PROVIDER SUMMARY

ElasticHosts is privately held it is the first public cloud service built on Linux Kernel-based Virtual Machine (rather than Xen), and it has an API that has been given a good review from a REST perspective. It is further unusual in that it charges by resources (CPU, Memory, Disk and Network) as separate entities. ElasticHosts makes flexible virtualization by allowing users to configure their sophisticated infrastructure to match thier business. It uses Linux KVM virtualization to provide operating-system-agnostic cloud servers.

Year Founded	2008	Moody's Rating	N/A	Pricing	Y

IaaS Sub-Classification

White Label	✓
Traditional	
Servers	✓
Storage	
Disaster Recovery	✓
Backup	✓
Messaging	
Content Delivery	
Desktop-as-a-Service	
Hosting	✓
IT-as-a-Service	
Management Software	

Direct & Channel

Peer1, CohesiveFT, OpenNebula.org, Linux

Clients

Trana eCommerce
North West Business Intelligence Ltd.
Au coin du jeu
Encircle Solutions Ltd.

Summary

Termination Notification (Days)	N/A
Dedicated Account Management	N/A
Live Customer Support	N/A
Guaranteed Network Availability	Y
Root/ Administrator Access	Y
Portal Support	N

Updates

Available
Q2 2011

CloudSourcing100.com

Regulation and Compliance

N/A

Programming Languages

N/A

Operating Systems Compatibility

CentOS 5.5, 5.4, 5.3; Debian 5.0; Knoppix 6.0.1; Red Hat Fedora 13, 12, 11, 10; Ubuntu 10.04, 9.10, 9.04, 8.10, 8.04, Windows Web Server 2008, Windows Web Server 2008 R2, Windows Server 2008 R2 Standard, Microsoft SQL Server 2008 Web Edition, FreeBSD 8.0, 7.2; OpenSolaris 2009.06

ALSBRIDGE

 **CLOUD** S O U R C I N G 100   **Q1 2011**

SOLUTION TYPE	PROVIDER TYPE | CUSTOMER CONTROL	
		where information lives® www.emc.com

PROVIDER SUMMARY

EMC Atmos Online is a Cloud Storage Service based on the EMC Atmos Storage Product. It is a reference architecture for customers that want to build a self-service storage cloud. Atmos Online demonstrates the ability to provide a multi-tenant, Internet-accessible storage resource that is infinitely scalable and designed for multiple EMC and third party applications, including solutions for Cloud backup (EMC Networker), Cloud archiving (EMC Celerra FAST), collaboration, content distribution and medical imaging.

Year Founded	1979	Moody's Rating	N/A	Pricing	N

IaaS Sub-Classification

White Label	
Traditional	✔
Servers	
Storage	✔
Disaster Recovery	
Backup	
Messaging	
Content Delivery	
Desktop-as-a-Service	
Hosting	
IT-as-a-Service	
Management Software	

Direct & Channel

Cirtas Bluejet Cloud, Riverbed, Oxygen Cloud, StorSimple, TwinStrata, Metalogix, CommVault, Seven10's StorFirst, Gladinet Cloud Desktop

Clients

ACUO Technologies
Atemp
Commvault
Metalogix
Signiant

Summary

Termination Notification (Days)	NG
Dedicated Account Management	Y
Live Customer Support	N
Guaranteed Network Availability	Y
Root/ Administrator Access	N
Portal Support	Y

Updates

Available
Q2 2011

CloudSourcing100.com

Regulation and Compliance

SAS70 Type II, PCI/DSS

Programming Languages

N/A

Operating Systems Compatibility

CentOS

ALSBRIDGE

CLOUD SOURCING 100

Q1 2011

SOLUTION TYPE	PROVIDER TYPE	CUSTOMER CONTROL	

www.engineyard.com

PROVIDER SUMMARY

Engine Yard AppCloud is an ideal Platform-as-a-Service for a range of Rails applications, from smaller-scale web applications that run within a single compute instance, to production applications that require the elasticity, scalability and reliability of a Rails Application Cloud. With Engine Yard AppCloud you only pay for what you use—there are no setup or environment fees, and you don't need your own Amazon account users get automated load balancing, persistent storage, web-based gem installs, data backup/restore, and system monitoring.

Year Founded	2006

Moody's Rating	N/A

Pricing	Y

IaaS Sub-Classification

White Label
Traditional
Servers
Storage
Disaster Recovery
Backup
Messaging
Content Delivery
Desktop-as-a-Service
Hosting
IT-as-a-Service
Management Software

Direct & Channel

Amazon Web Services, Terremark, Send-Grid, New Relic

Clients

MTV
EarthAid Enterprises
Ellison
Path
Bayer

Summary

Termination Notification (Days)	N/A
Dedicated Account Management	N/A
Live Customer Support	Y
Guaranteed Network Availability	Y
Root/ Administrator Access	Y
Portal Support	Y

Updates

Available
Q2 2011

CloudSourcing100.com

Regulation and Compliance

N/A

Operating Systems Compatibility

Linux Operating Systems, Windows Server 2008

Programming Languages

Ruby

 **Q1 2011**

SOLUTION TYPE

PROVIDER TYPE | CUSTOMER CONTROL

www.enki.co

PROVIDER SUMMARY

ENKI provides multiple cloud-based solutions that are designed to assist companies with reducing the costs and distractions associated with managing IT. ENKI Virtual IT provides users with full, on-demand management of thier application lifecycle, from development to production. ENKI solutions offer both ITaaS and IaaS capabilities through their PrimaCare and PrimaCloud offerings.

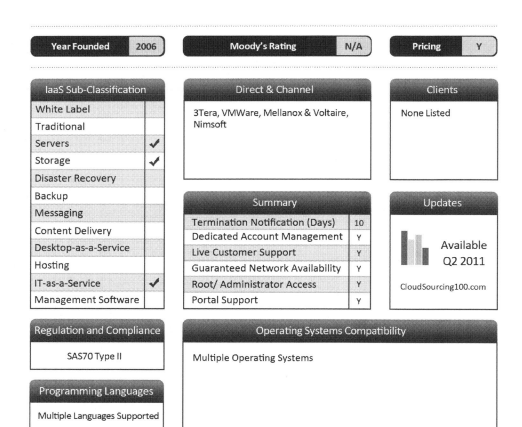

Year Founded	2006

Moody's Rating	N/A

Pricing	Y

IaaS Sub-Classification

White Label	
Traditional	
Servers	✓
Storage	✓
Disaster Recovery	
Backup	
Messaging	
Content Delivery	
Desktop-as-a-Service	
Hosting	
IT-as-a-Service	✓
Management Software	

Regulation and Compliance

SAS70 Type II

Programming Languages

Multiple Languages Supported

Direct & Channel

3Tera, VMWare, Mellanox & Voltaire, Nimsoft

Summary

Termination Notification (Days)	10
Dedicated Account Management	Y
Live Customer Support	Y
Guaranteed Network Availability	Y
Root/ Administrator Access	Y
Portal Support	Y

Operating Systems Compatibility

Multiple Operating Systems

Clients

None Listed

Updates

Available
Q2 2011

CloudSourcing100.com

CLOUD SOURCING 100

Q1 2011

SOLUTION TYPE	PROVIDER TYPE | CUSTOMER CONTROL	

www.enomaly.com

PROVIDER SUMMARY

As one of the earliest pioneers of cloud computing, Enomaly has the experience to monetize the cloud. Enomaly ECP 3 technology, has benefited from 6+ years of cloud computing leadership, and used by telecom and IDC operators in North America, the UK, Europe, and Asia to deliver cloud computing services to their customers. The Elastic Computing Platform (ECP) Version 3 is a full featured cloud computing environment for service providers and Internet Data Centers looking to offer revenue generating cloud services. ECP has been designed to meet the most rigorous of IT demands while also remaining easy to administer and use.

Year Founded	2004

Moody's Rating	N/A

Pricing	N

IaaS Sub-Classification

White Label	
Traditional	
Servers	
Storage	
Disaster Recovery	
Backup	
Messaging	
Content Delivery	
Desktop-as-a-Service	
Hosting	
IT-as-a-Service	
Management Software	✔

Direct & Channel

Intel, HP, RedHat, CloudTest, Ericom, Gemniare

Clients

Bank of China
City Cloud
CentriLogic
Hosts Unlimited
Changemakers

Summary

Termination Notification (Days)	N/A
Dedicated Account Management	Y
Live Customer Support	N/A
Guaranteed Network Availability	N/A
Root/ Administrator Access	Y
Portal Support	N/A

Updates

Available
Q2 2011

CloudSourcing100.com

Regulation and Compliance

N/A

Operating Systems Compatibility

N/A

Programming Languages

N/A

 CLOUD SOURCING 100 **Q1 2011**

SOLUTION TYPE   PROVIDER TYPE | CUSTOMER CONTROL

www.enstratus.com

PROVIDER SUMMARY

enStratus™ is a cloud infrastructure management solution for deploying and managing enterprise-class applications in public, private and hybrid clouds. The company defines cloud governance to mean: Security controls, including user management, encryption, and key management, Financial controls, including cloud cost tracking, budgets, chargebacks, and multi-currency support, Audit controls and reporting, Monitoring and alerting, Automation, including auto-scaling, cloud bursting, backup management, and change management, Unified cross-cloud management. enStratus supports both SaaS and on-premise deployment models.

Year Founded	2008		Moody's Rating	N/A		Pricing	N

IaaS Sub-Classification

White Label	
Traditional	
Servers	
Storage	
Disaster Recovery	
Backup	
Messaging	
Content Delivery	
Desktop-as-a-Service	
Hosting	
IT-as-a-Service	
Management Software	✓

Direct & Channel

Amazon Web Services, Windows Azure, Rackspace, GoGrid, ReliaCloud, Eucalyptus

Summary

Termination Notification (Days)	N/A
Dedicated Account Management	N/A
Live Customer Support	N/A
Guaranteed Network Availability	N/A
Root/ Administrator Access	Y
Portal Support	Y

Clients

KT
8th Bridge
Quantum Retail
CSA
SAIC

Updates

Available
Q2 2011

CloudSourcing100.com

Regulation and Compliance

N/A

Operating Systems Compatibility

Ubuntu, Debian, CentOS, Fedora, Red Hat, Solaris, Windows 2003 and Windows 2008

Programming Languages

Bash, Python

 **Q1 2011**

SOLUTION TYPE	PROVIDER TYPE | CUSTOMER CONTROL	

www.flexiant.com

PROVIDER SUMMARY

Flexiant is a provider of cloud computing infrastructure, both as a public platform (FlexiScale) and as a licensed product for data center owners (Extility). FlexiScale provides a wholly scalable hosting infrastructure and also reaches out to the wider world of IT services delivery companies who, as hosting resellers, have the opportunity to extend their offering to their customers. Extility enables hosting companies to compete in the cloud computing virtualization sector on a level playing field with global market leaders who have proprietary rapid provisioning infrastructures.

Year Founded	2009		Moody's Rating	N/A		Pricing	Y

IaaS Sub-Classification

White Label	
Traditional	
Servers	✔
Storage	
Disaster Recovery	
Backup	
Messaging	
Content Delivery	
Desktop-as-a-Service	
Hosting	
IT-as-a-Service	
Management Software	

Direct & Channel

CohesiveFT, RightScale

Clients

None Listed

Summary

Termination Notification (Days)	N/A
Dedicated Account Management	N
Live Customer Support	Y
Guaranteed Network Availability	Y
Root/ Administrator Access	Y
Portal Support	Y

Updates

Available Q2 2011

CloudSourcing100.com

Regulation and Compliance

N/A

Programming Languages

Multiple Languages Supported

Operating Systems Compatibility

Windows Server 2008, CentOS Linux 5.4, Debian Linux 5.0, Ubuntu Linux 9.10, Ubuntu Linux 10.04 LTS, and more in the works. Customers can install their own operating systems (32 bit or 64 bit) by downloading and booting from the appropriate ISO image.

CLOUD SOURCING 100 — Q1 2011

SOLUTION TYPE	PROVIDER TYPE | CUSTOMER CONTROL

Force.com
www.force.com

PROVIDER SUMMARY

Force.com offers a scalable, and secure platform for application development. It delivers a complete technology stack from database and security to workflow and user interface—so you can focus on assembling, building, and instantly deploying solutions. As a result, custom application development is possible without the expense of buying, configuring, and managing development hardware and software. Force.com speeds innovation through a powerful yet easy-to-use application development and deployment model. You can easily develop and then immediately deploy your solutions to the Force.com cloud-based infrastructure.

Year Founded	1999		Moody's Rating	N/A		Pricing	Y

IaaS Sub-Classification	
White Label	
Traditional	
Servers	
Storage	
Disaster Recovery	
Backup	
Messaging	
Content Delivery	
Desktop-as-a-Service	
Hosting	
IT-as-a-Service	
Management Software	

Direct & Channel
3000+ successful partners and counting

Clients
Mulitple Client Listings

Summary	
Termination Notification (Days)	N/A
Dedicated Account Management	Y
Live Customer Support	Y
Guaranteed Network Availability	Y
Root/ Administrator Access	N
Portal Support	Y

Updates
Available Q2 2011
CloudSourcing100.com

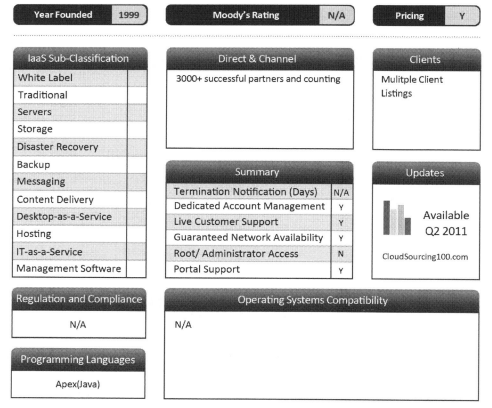

Regulation and Compliance
N/A

Operating Systems Compatibility
N/A

Programming Languages
Apex(Java)

ALSBRIDGE

CLOUD SOURCING 100

Q1 2011

SOLUTION TYPE

PROVIDER TYPE | CUSTOMER CONTROL

www.genpact.com

PROVIDER SUMMARY

Genpact is a global IT Infrastructure services provider that plans, designs and implements organizational IT strategies and manages mission-critical IT Infrastructure for global clients. Genpact leverages its in-depth industry and technical knowledge and helps clients extract maximum value from their IT investments. The company delivers this by providing an IT Infrastructure management platform that is cost effective, reliable and cutting edge.

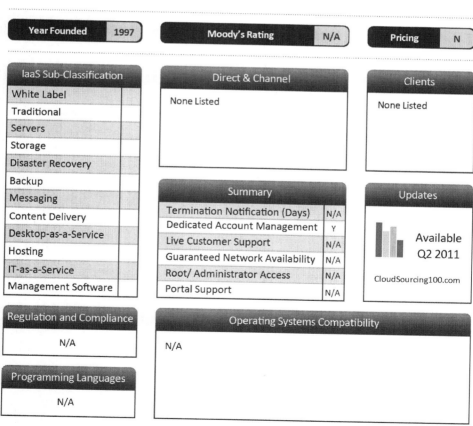

Year Founded	1997

Moody's Rating	N/A

Pricing	N

IaaS Sub-Classification

White Label	
Traditional	
Servers	
Storage	
Disaster Recovery	
Backup	
Messaging	
Content Delivery	
Desktop-as-a-Service	
Hosting	
IT-as-a-Service	
Management Software	

Direct & Channel

None Listed

Clients

None Listed

Summary

Termination Notification (Days)	N/A
Dedicated Account Management	Y
Live Customer Support	N/A
Guaranteed Network Availability	N/A
Root/ Administrator Access	N/A
Portal Support	N/A

Updates

Available
Q2 2011

CloudSourcing100.com

Regulation and Compliance

N/A

Programming Languages

N/A

Operating Systems Compatibility

N/A

ALSBRIDGE

CLOUD  SOURCING 100   Q1 2011

SOLUTION TYPE   PROVIDER TYPE | CUSTOMER CONTROL

www.gigaspaces.com

PROVIDER SUMMARY

GigaSpaces Technologies is a provider of a new generation of application platforms for Java and .Net environments that offer an alternative to traditional application-servers. The company's eXtreme Application Platform (XAP) is a high-end application server, designed to meet the most demanding business requirements. It is the product that provides a complete middleware solution on a scalable platform. GigaSpaces has developed its 2nd-Generation PaaS Enablement platform built from the ground up to provide a complete, silo-free, and agile service development and operational environment that facilitates application migration to the cloud.

Year Founded	2000		Moody's Rating	N/A		Pricing	N

IaaS Sub-Classification

White Label	
Traditional	
Servers	
Storage	
Disaster Recovery	
Backup	
Messaging	
Content Delivery	
Desktop-as-a-Service	
Hosting	
IT-as-a-Service	
Management Software	

Direct & Channel

HP Business Partner, Nortel, Amazon Web Services, Integrasoft, Intel, Microsoft, SGI and many more

Clients

Chartwell
First Data
Gallup
JDSU
Sears Holdings

Summary

Termination Notification (Days)	N/A
Dedicated Account Management	N/A
Live Customer Support	Y
Guaranteed Network Availability	N/A
Root/ Administrator Access	Y
Portal Support	Y

Updates

Available
Q2 2011

CloudSourcing100.com

Regulation and Compliance

N/A

Operating Systems Compatibility

N/A

Programming Languages

Java, .NET

CLOUD SOURCING 100   **Q1 2011**

SOLUTION TYPE	PROVIDER TYPE | CUSTOMER CONTROL	

www.gogrid.com

PROVIDER SUMMARY

GoGrid is a cloud infrastructure service, hosting Linux and Windows virtual machines managed by a multi-server control panel. GoGrid cloud hosting allows users to build scalable cloud infrastructure in multiple data centers using cloud servers, elastic F5 hardware load balancing, and cloud storage with total control through automation and self-service. GoGrid is a pure-play Infrastructure-as-a-Service (IaaS) provider specializing in Cloud Infrastructure solutions. They make complex infrastructure easy by enabling businesses to revolutionize their IT environments with the Cloud.

Year Founded	2001

Moody's Rating	N/A

Pricing	Y

IaaS Sub-Classification

White Label	✔
Traditional	
Servers	✔
Storage	✔
Disaster Recovery	
Backup	
Messaging	
Content Delivery	✔
Desktop-as-a-Service	
Hosting	
IT-as-a-Service	
Management Software	

Direct & Channel

BitNami LAMPstack, Moodle, cPanel, CohesiveFT VPN Cubed Ipsec

Clients

None Listed

Summary

Termination Notification (Days)	1
Dedicated Account Management	Y
Live Customer Support	Y
Guaranteed Network Availability	Y
Root/ Administrator Access	Y
Portal Support	Y

Updates

Available Q2 2011

CloudSourcing100.com

Regulation and Compliance

SAS70 Type II

Programming Languages

Multiple Languages Supported

Operating Systems Compatibility

Cent OS 5.1, Red Hat 5.1, Windows Server 2003, Windows Server 2008

ALSBRIDGE

 **Q1 2011**

SOLUTION TYPE	PROVIDER TYPE	CUSTOMER CONTROL	

www.google.com/apps

PROVIDER SUMMARY

Google App Engine is a platform for developing and hosting web applications in Google-managed data centers. It virtualizes applications across multiple servers and data centers. App Engine for Business provides all the ease of use and flexibility of App Engine with more power to manage enterprise use cases, more capable APIs, straightforward pricing and the SLAs and support business-critical applications.

Year Founded	2008		Moody's Rating	Aa2		Pricing	Y

IaaS Sub-Classification
White Label
Traditional
Servers
Storage
Disaster Recovery
Backup
Messaging
Content Delivery
Desktop-as-a-Service
Hosting
IT-as-a-Service
Management Software

Direct & Channel
Multiple Partner Listings

Clients
Genentech
Virgin America
National Geographic
Emerson
Nimble Books

Summary	
Termination Notification (Days)	N/A
Dedicated Account Management	Y
Live Customer Support	Y
Guaranteed Network Availability	Y
Root/ Administrator Access	N
Portal Support	Y

Updates
Available Q2 2011
CloudSourcing100.com

Regulation and Compliance
European Union Safe Harbour

Operating Systems Compatibility
Linux Operating Systems

Programming Languages
Java, Python

ALSBRIDGE

CLOUD SOURCING 100

Q1 2011

SOLUTION TYPE

PROVIDER TYPE | CUSTOMER CONTROL

GridGain

www.gridgain.com

PROVIDER SUMMARY

GridGain provides state of the art implementations for computational grids, data grids, and cloud auto-scaling - all built based on a unique and patent-pending zero deployment technology. You can develop with GridGain using Java or Scala programming languages - both of which are supported natively. GridGain provides many features that make development of highly scalable distributed applications easier and productive.

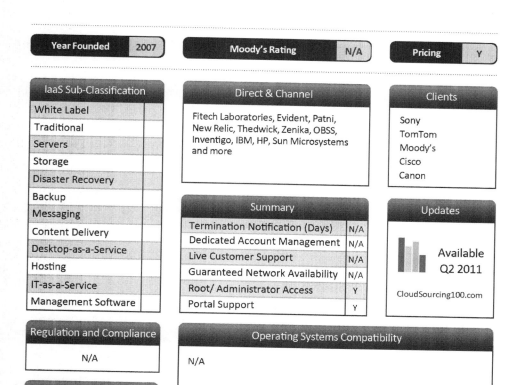

Year Founded	2007
Moody's Rating	N/A
Pricing	Y

IaaS Sub-Classification

- White Label
- Traditional
- Servers
- Storage
- Disaster Recovery
- Backup
- Messaging
- Content Delivery
- Desktop-as-a-Service
- Hosting
- IT-as-a-Service
- Management Software

Direct & Channel

Fitech Laboratories, Evident, Patni, New Relic, Thedwick, Zenika, OBSS, Inventigo, IBM, HP, Sun Microsystems and more

Clients

Sony
TomTom
Moody's
Cisco
Canon

Summary

Termination Notification (Days)	N/A
Dedicated Account Management	N/A
Live Customer Support	N/A
Guaranteed Network Availability	N/A
Root/ Administrator Access	Y
Portal Support	Y

Updates

Available Q2 2011

CloudSourcing100.com

Regulation and Compliance

N/A

Operating Systems Compatibility

N/A

Programming Languages

JAVA, Scala

ALSBRIDGE

CLOUD SOURCING 100   Q1 2011

SOLUTION TYPE	PROVIDER TYPE | CUSTOMER CONTROL

          Hewlett-Packard

www.hp.com

PROVIDER SUMMARY

HP CloudSystem is built on the HP Cloud Service Automaton and Converged Infrastructure tech-nologies. With support a broad set of applications, CloudSystem provides users with a unifed way to offer, provision and manage services across private clouds, public cloud providers, and tradi-tional IT. It enables users the flexibility to scale capacity within and outside their data center. CloudSystem is extensible to users existing IT infrastructure and can support heterogeneous envi-ronments.

Year Founded	1939		Moody's Rating	A2		Pricing	N

IaaS Sub-Classification

White Label	
Traditional	✓
Servers	✓
Storage	
Disaster Recovery	
Backup	
Messaging	
Content Delivery	
Desktop-as-a-Service	
Hosting	
IT-as-a-Service	
Management Software	

Direct & Channel

None Specific to Cloud

Clients

None Specific to Cloud

Summary

Termination Notification (Days)	N/A
Dedicated Account Management	Y
Live Customer Support	Y
Guaranteed Network Availability	Y
Root/ Administrator Access	N
Portal Support	N/A

Updates

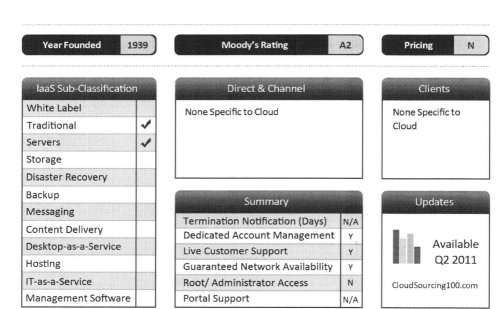

Available Q2 2011

CloudSourcing100.com

Regulation and Compliance

N/A

Operating Systems Compatibility

Cent OS, Windows Server 2003, Windows Server 2008

Programming Languages

Java, Ruby, SQL

CLOUD SOURCING 100

Q1 2011

SOLUTION TYPE	PROVIDER TYPE | CUSTOMER CONTROL	
		www.huawei.com

PROVIDER SUMMARY

Huawei is a provider of next-generation telecommunications network solutions for operators around the world. Its Cloud Computing strategy for the North American market is designed to advance customers' journey to cloud computing, and involves working with leading North American partners to deliver solutions that enable customers to expand service offerings, reduce costs, and bring telecommunications services to the cloud.

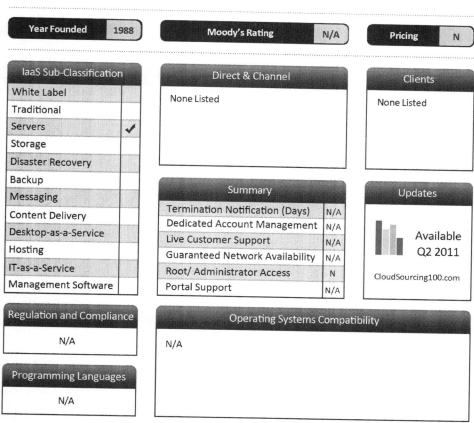

Year Founded	1988

Moody's Rating	N/A

Pricing	N

IaaS Sub-Classification

White Label	
Traditional	
Servers	✓
Storage	
Disaster Recovery	
Backup	
Messaging	
Content Delivery	
Desktop-as-a-Service	
Hosting	
IT-as-a-Service	
Management Software	

Direct & Channel

None Listed

Clients

None Listed

Summary

Termination Notification (Days)	N/A
Dedicated Account Management	N/A
Live Customer Support	N/A
Guaranteed Network Availability	N/A
Root/ Administrator Access	N
Portal Support	N/A

Updates

Available Q2 2011

CloudSourcing100.com

Regulation and Compliance

N/A

Operating Systems Compatibility

N/A

Programming Languages

N/A

ALSBRIDGE

 **Q1 2011**

SOLUTION TYPE

PROVIDER TYPE | CUSTOMER CONTROL

Hyperic
www.hyperic.com

PROVIDER SUMMARY

Hyperic's CloudStatus is built on Hyperic HQ, Hyperic's flagship product designed to monitor and manage large scale web infrastructure. The Hyperic HQ Server aggregates multiple metrics from sources inside and outside the cloud to provide cloud availability and health status. Hyperic HQ then calculates the aggregate data to determine overall availability and normalized metrics across the cloud. The multiple metrics origination scheme assures users a relevant overall perspective of cloud performance.

Year Founded	2004

Moody's Rating	N/A

Pricing	N

IaaS Sub-Classification

White Label	
Traditional	
Servers	
Storage	
Disaster Recovery	
Backup	
Messaging	
Content Delivery	
Desktop-as-a-Service	
Hosting	
IT-as-a-Service	
Management Software	✓

Direct & Channel

Sopera, openNMS, Unisys, VMWare, IBM, Microsoft, Citrix, Hewlett-Packard

Clients

Mosso
BCM
Contegix
Ogilvy
MySql

Summary

Termination Notification (Days)	N/A
Dedicated Account Management	N/A
Live Customer Support	Y
Guaranteed Network Availability	N/A
Root/ Administrator Access	N
Portal Support	Y

Updates

Available Q2 2011

CloudSourcing100.com

Regulation and Compliance

N/A

Operating Systems Compatibility

N/A

Programming Languages

N/A

CLOUD SOURCING 100

Q1 2011

SOLUTION TYPE	PROVIDER TYPE | CUSTOMER CONTROL

IBM

www.ibm.com

PROVIDER SUMMARY

IBM offers a host of services such as IBM Strategy and Change Services for Cloud Adoption and IBM Strategy and Design Services for a Cloud Infrastructure to help clients develop a cloud road-map. IBM's enterprise cloud computing services enable users to assess cloud readiness, develop adoption strategies and identify business entry points. Their security and experience go into each of their global computer center and myriad enterprise private clouds resulting in an instrument-ed, interconnected, intelligent approach to smarter computing

Year Founded	1911

Moody's Rating	Aa3

Pricing	N

IaaS Sub-Classification

White Label	
Traditional	✓
Servers	✓
Storage	✓
Disaster Recovery	✓
Backup	
Messaging	✓
Content Delivery	
Desktop-as-a-Service	✓
Hosting	✓
IT-as-a-Service	✓
Management Software	

Direct & Channel

None Specific to Cloud

Clients

NedBank
North Carolina State
Wuxi Lake Tai
Getronics
BlueLock

Summary

Termination Notification (Days)	NG
Dedicated Account Management	y
Live Customer Support	y
Guaranteed Network Availability	y
Root/ Administrator Access	Y
Portal Support	Y

Updates

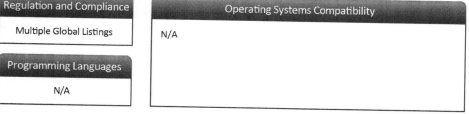

Available
Q2 2011

CloudSourcing100.com

Regulation and Compliance

Multiple Global Listings

Programming Languages

N/A

Operating Systems Compatibility

N/A

ALSBRIDGE

 **Q1 2011**

SOLUTION TYPE	PROVIDER TYPE | CUSTOMER CONTROL	

IBM (BPaaS)
www.ibm.com

PROVIDER SUMMARY

IBM provides services to decide which of the new cloud-based delivery models should be considered based on existing IT and business strategies: BPaaS (Business Process as a Service)- Business process services, such as billing, contract management, payroll, HR and cloud-based advertising , SaaS (Software as a Service)–Standardized, network-delivered IT applications, including CRM, collaboration and analytics programs ,PaaS (Platform as a Service)–Application development environments ,IaaS (Infrastructure as a Service)–Infrastructure needs, including server and storage computing power.

Year Founded	1911		Moody's Rating	Aa3		Pricing	N

IaaS Sub-Classification
White Label
Traditional
Servers
Storage
Disaster Recovery
Backup
Messaging
Content Delivery
Desktop-as-a-Service
Hosting
IT-as-a-Service
Management Software

Direct & Channel
None Specific to Cloud

Clients
NedBank
North Carolina State
Wuxi Lake Tai
Getronics
BlueLock

Summary
Termination Notification (Days)	NG
Dedicated Account Management	Y
Live Customer Support	N/A
Guaranteed Network Availability	N/A
Root/ Administrator Access	N/A
Portal Support	N/A

Updates

Available
Q2 2011

CloudSourcing100.com

Regulation and Compliance
Multiple Global Listings

Operating Systems Compatibility
N/A

Programming Languages
N/A

CLOUD SOURCING 100   Q1 2011

SOLUTION TYPE

PROVIDER TYPE | CUSTOMER CONTROL

www.icloud.com

PROVIDER SUMMARY

iCloud is an online computer, like an operating system running in the cloud with an AJAX-based remote web desktop. It is accessible from both desktop operating systems and mobile operating systems using WebDAV. Because it's running in the cloud (the internet) it can offer you impressive features such as easy sharing and rich collaboration.

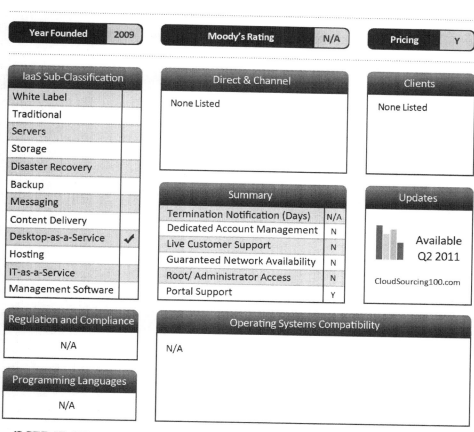

| Year Founded | 2009 | | Moody's Rating | N/A | | Pricing | Y |

IaaS Sub-Classification

White Label	
Traditional	
Servers	
Storage	
Disaster Recovery	
Backup	
Messaging	
Content Delivery	
Desktop-as-a-Service	✔
Hosting	
IT-as-a-Service	
Management Software	

Direct & Channel

None Listed

Clients

None Listed

Summary

Termination Notification (Days)	N/A
Dedicated Account Management	N
Live Customer Support	N
Guaranteed Network Availability	N
Root/ Administrator Access	N
Portal Support	Y

Updates

Available
Q2 2011

CloudSourcing100.com

Regulation and Compliance

N/A

Operating Systems Compatibility

N/A

Programming Languages

N/A

ALSBRIDGE

CLOUD SOURCING 100

Q1 2011

SOLUTION TYPE	PROVIDER TYPE	CUSTOMER CONTROL

www.internap.com

PROVIDER SUMMARY

Internap provides high performance, highly available storage solution that can reduce your costs while improving the end-user experience. It's cloud, only faster. Internap's XIPCloud™ Storage solution provides users with storage located in secured data centers and delivered over Internap's Performance IP™ service, Internap XIPCloud Storage offers unlimited storage capacity, on-demand access, and utility billing. The service provides an extensive, open API and a variety of tools to manage your data.

Year Founded	1996		Moody's Rating	N/A		Pricing	N

IaaS Sub-Classification	
White Label	
Traditional	
Servers	
Storage	
Disaster Recovery	
Backup	
Messaging	
Content Delivery	✔
Desktop-as-a-Service	
Hosting	✔
IT-as-a-Service	
Management Software	

Direct & Channel

None Listed

Clients

Carbonite
Digital Globe
Apex Learning
Sundance Institute
Online

Summary	
Termination Notification (Days)	N/A
Dedicated Account Management	N/A
Live Customer Support	Y
Guaranteed Network Availability	Y
Root/ Administrator Access	N
Portal Support	Y

Updates

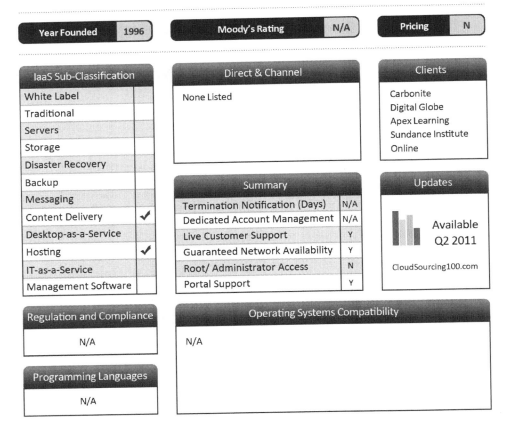

Available
Q2 2011

CloudSourcing100.com

Regulation and Compliance

N/A

Operating Systems Compatibility

N/A

Programming Languages

N/A

CLOUD SOURCING 100 — Q1 2011

SOLUTION TYPE	PROVIDER TYPE | CUSTOMER CONTROL	

www.ironmountain.com

PROVIDER SUMMARY

Iron Mountain offers multiple solutions, from Backup as a Service to Storage to Disaster Recovery, Health Information Management, Software as a Service Escrow for document management and archiving across mulitple industries. The best known Iron Mountain storage facility is a high-security storage facility in a former limestone mine at Boyers, Pennsylvania near the city of Butler in the United States. Iron Mountain has additional underground storage facilities in the United States and the rest of the world. Most of the company's over 1,000 storage locations are in above-ground leased warehouse space located near customers.

Year Founded	1951	Moody's Rating	Ba3	Pricing	N

IaaS Sub-Classification

White Label	
Traditional	✔
Servers	
Storage	✔
Disaster Recovery	✔
Backup	✔
Messaging	
Content Delivery	
Desktop-as-a-Service	
Hosting	
IT-as-a-Service	
Management Software	

Direct & Channel

AHR Consulting, Data Mountain, Dataline, ePartners, LMC Data, Synegi, Compsat, Dallas Digital

Summary

Termination Notification (Days)	N/A
Dedicated Account Management	N/A
Live Customer Support	Y
Guaranteed Network Availability	N/A
Root/ Administrator Access	N
Portal Support	N

Clients

Multiple Client Listings

Updates

Available Q2 2011

CloudSourcing100.com

Regulation and Compliance

SysTrust, PCI/DSS Level 1, AAA certified

Programming Languages

N/A

Operating Systems Compatibility

N/A

 CLOUD SOURCING 100 **Q1 2011**

SOLUTION TYPE   PROVIDER TYPE | CUSTOMER CONTROL

www.joyentcloud.com

PROVIDER SUMMARY

Joyent is a global cloud computing software company offering cloud computing solutions world-wide. Joyent licenses its cloud software, SmartDataCenter, to communications service providers, like Dell and China Cache, who deliver cloud services to their own customers. In addition, Joyent runs an instantiation of SmartDataCenter for some of the most innovative companies in the world, such as LinkedIn, Gilt Groupe and Kabam. Joyent's Smart Technologies provide application virtualization. Their approach provides applications with better performance and cost effectiveness, due to higher application density and hardware utilization in their Smart Technology stack.

Year Founded	2004		Moody's Rating	N/A		Pricing	Y

IaaS Sub-Classification

White Label	
Traditional	
Servers	✔
Storage	
Disaster Recovery	
Backup	
Messaging	
Content Delivery	
Desktop-as-a-Service	
Hosting	
IT-as-a-Service	
Management Software	

Direct & Channel

Intel, DELL, Arista, Switch, Riak, Zeus, MySQL, Guardtime, Tmpsocial, Code-sion, New Relic

Clients

GILT Groupe
LinkedIn
KABAM
THQ
AKQA

Summary

Termination Notification (Days)	0
Dedicated Account Management	Y
Live Customer Support	Y
Guaranteed Network Availability	Y
Root/ Administrator Access	Y
Portal Support	Y

Updates

 Available Q2 2011

CloudSourcing100.com

Regulation and Compliance

SAS70 Type II

Operating Systems Compatibility

Open Solaris

Programming Languages

PHP, Java, Python, Erlang, C, C++, Ruby, and more

 ALSBRIDGE

CLOUD SOURCING 100   Q1 2011

SOLUTION TYPE

   B

PROVIDER TYPE | CUSTOMER CONTROL

 F   H

JumpBox
www.jumpbox.com

PROVIDER SUMMARY

JumpBox is a server software management company. JumpBox is focused on easing the deployment and management of various different software for collaborative projects. A JumpBox is a "ready-to-use" virtual machine that will run on any computing environment that supports virtualization. A single download is compatible with all major forms of virtualization like VMware, Parallels, Microsoft, Xen Open Source, VirtualBox and Amazon EC2. The JumpBox library contains more than fifty-five different applications spanning all major product categories.

Year Founded	2006

Moody's Rating	N/A

Pricing	Y

IaaS Sub-Classification	
White Label	
Traditional	
Servers	
Storage	
Disaster Recovery	
Backup	
Messaging	
Content Delivery	
Desktop-as-a-Service	
Hosting	
IT-as-a-Service	
Management Software	

Direct & Channel
IBM, OnApp, VMWare, Parallels, ubuntu, Citrix, Zenoss

Clients
Brandup
Fabio Perini
SimulTrans
HJ Harkins

Summary	
Termination Notification (Days)	N/A
Dedicated Account Management	N/A
Live Customer Support	Y
Guaranteed Network Availability	N/A
Root/ Administrator Access	Y
Portal Support	Y

Updates
Available Q2 2011
CloudSourcing100.com

Regulation and Compliance
N/A

Operating Systems Compatibility
N/A

Programming Languages
N/A

ALSBRIDGE

© Alsbridge Inc, 2011

 **CLOUD** SOURCING 100   **Q1 2011**

SOLUTION TYPE	PROVIDER TYPE | CUSTOMER CONTROL	
		Kaavo www.kaavo.com

PROVIDER SUMMARY

Kaavo provides solutions for deploying and managing on-demand applications and workload in the cloud. It provides application-centric management of cloud resources, framework to automate the deployment and run-time management (production support) of applications and workloads on the cloud. Public and private cloud providers as well as enterprise customers use Kaavo's technology to deliver solutions for managing distributed applications and workloads in public, private, and hybrid clouds.

Year Founded	2007		Moody's Rating	N/A		Pricing	Y

IaaS Sub-Classification

White Label	
Traditional	
Servers	
Storage	
Disaster Recovery	
Backup	
Messaging	
Content Delivery	
Desktop-as-a-Service	
Hosting	
IT-as-a-Service	
Management Software	✔

Direct & Channel

IBM, Amazon Web Services, Rackspace, NIIT Technologies, CITRIX, VMWare

Clients

CapCal
SellPoint

Summary

Termination Notification (Days)	N/A
Dedicated Account Management	N
Live Customer Support	N
Guaranteed Network Availability	N/A
Root/ Administrator Access	Y
Portal Support	Y

Updates

Available
Q2 2011

CloudSourcing100.com

Regulation and Compliance

N/A

Operating Systems Compatibility

N/A

Programming Languages

N/A

CLOUD SOURCING 100

Q1 2011

SOLUTION TYPE	PROVIDER TYPE | CUSTOMER CONTROL	

www.layer7tech.com

PROVIDER SUMMARY

Layer 7 Technologies provides API security and governance for service-oriented, Web-oriented and cloud-oriented integration. Through their line of SecureSpan and CloudSpan family of API gateways and management products, Layer 7 enables organizations to control how they expose their data and applications to outside divisions, partners, mobile developers and cloud services. The Layer 7 SecureSpan XML Gateway provides a wide range of capabilities for SOA, web services and cloud environments, such as regulating access to services, enforcing policies, protecting against malicious attacks and enabling policy-based integration.

Year Founded	2003

Moody's Rating	N/A

Pricing	N

IaaS Sub-Classification

White Label	
Traditional	
Servers	
Storage	
Disaster Recovery	
Backup	
Messaging	
Content Delivery	
Desktop-as-a-Service	
Hosting	
IT-as-a-Service	
Management Software	✔

Direct & Channel

Oracle, Sun, SoftwareAG, Cisco, Novella and more

Clients

BAE Systems
Chrysler Financial
Aviall
IBM
Dell

Summary

Termination Notification (Days)	N/A
Dedicated Account Management	Y
Live Customer Support	Y
Guaranteed Network Availability	N/A
Root/ Administrator Access	Y
Portal Support	Y

Updates

Available
Q2 2011

CloudSourcing100.com

Regulation and Compliance

N/A

Programming Languages

N/A

Operating Systems Compatibility

N/A

ALSBRIDGE

CLOUD SOURCING 100

Q1 2011

SOLUTION TYPE

PROVIDER TYPE | CUSTOMER CONTROL

www.layeredtech.com

PROVIDER SUMMARY

Layered Tech is a global provider of managed dedicated hosting, on-demand grid/virtualization computing and Web services. Their innovative solutions let customers streamline and fully utilize their server resources. With seven top-tier data centers around the world, their infrastructure powers millions of sites and Internet-enabled applications including e-commerce, software as a service (SaaS) and content distribution.

Year Founded	2004

Moody's Rating	N/A

Pricing	Y

IaaS Sub-Classification

White Label	
Traditional	
Servers	✔
Storage	✔
Disaster Recovery	
Backup	
Messaging	
Content Delivery	
Desktop-as-a-Service	
Hosting	✔
IT-as-a-Service	
Management Software	

Direct & Channel

Microsoft, VMWare Cisco, 3Tera, DELL, Internap

Summary

Termination Notification (Days)	2
Dedicated Account Management	N/A
Live Customer Support	Y
Guaranteed Network Availability	Y
Root/ Administrator Access	Y
Portal Support	Y

Clients

KANA Software
TomGreen
SilkFair
1Dawg
MobileStorm

Updates

Available
Q2 2011

CloudSourcing100.com

Regulation and Compliance

PCI/DSS

Programming Languages

N/A

Operating Systems Compatibility

CentOS 4.x and above, Red Hat Enterprise Linux (RHEL) 4.x, Windows Server 2008 R2, Windows Server 2008

ALSBRIDGE

CLOUD SOURCING 100 Q1 2011

SOLUTION TYPE

PROVIDER TYPE | CUSTOMER CONTROL

LocalMirror
www.localmirror.com

PROVIDER SUMMARY

LocalMirror CDN technology distributes file downloads and audio/video streams from the closest location & with lower latency, thus offering better Internet experience for your users compared with traditional hosting & streaming. The service can be easily activated for static content in a matter of hours and offload server load & internet connectivity traffic as content will be served from LocalMirror CDN nodes that are dispersed around the world. They provide thousands of concurrent connections & high bandwidth download service for web sites & on-line portals that require electronic software or any other static content distribution via HTTP & FTP protocols.

Year Founded	N/A

Moody's Rating	N/A

Pricing	N

IaaS Sub-Classification

White Label	
Traditional	
Servers	
Storage	
Disaster Recovery	
Backup	
Messaging	
Content Delivery	✔
Desktop-as-a-Service	
Hosting	
IT-as-a-Service	
Management Software	

Direct & Channel

None Listed

Clients

None Listed

Summary

Termination Notification (Days)	N/A
Dedicated Account Management	N/A
Live Customer Support	N
Guaranteed Network Availability	Y
Root/ Administrator Access	N
Portal Support	Y

Updates

Available
Q2 2011

CloudSourcing100.com

Regulation and Compliance

N/A

Operating Systems Compatibility

N/A

Programming Languages

N/A

ALSBRIDGE

 **Q1 2011**

SOLUTION TYPE

PROVIDER TYPE | CUSTOMER CONTROL

Longjump
www.longjump.com

PROVIDER SUMMARY

LongJump has been in cloud application technologies since 2003 with one of the first multitenant SaaS platforms for creating business applications. LongJump provides a platform centered around a development methodology that scales and adapts to changing market conditions. LongJump began with the following belief: Precious development resources are better utilized creating innovative solutions, rather than re-inventing common technologies. It's primary objective is to eliminate the cost associated with engineering and architecting reinvention efforts involved in bringing multitenant SaaS applications to market.

Year Founded	2003		Moody's Rating	N/A		Pricing	Y

IaaS Sub-Classification

White Label	
Traditional	
Servers	
Storage	
Disaster Recovery	
Backup	
Messaging	
Content Delivery	
Desktop-as-a-Service	
Hosting	
IT-as-a-Service	
Management Software	

Regulation and Compliance

SAS70 Type II

Programming Languages

Java

Direct & Channel

AscendWorks, Aspire Systems, BMA, FA-onDemand, Force by Design, Ideabox, jamcracker, Scio

Summary

Termination Notification (Days)	5
Dedicated Account Management	N/A
Live Customer Support	N/A
Guaranteed Network Availability	N/A
Root/ Administrator Access	Y
Portal Support	Y

Operating Systems Compatibility

CentOS

Clients

Nielson
Simco Electronics
Media News Group
Hasselblad
DavidAllen

Updates

 Available Q2 2011

CloudSourcing100.com

CLOUD SOURCING 100   Q1 2011

SOLUTION TYPE	PROVIDER TYPE	CUSTOMER CONTROL	
			 www.mezeo.com

PROVIDER SUMMARY

Mezeo Software provides a deployable, software-only REST API accessible cloud storage platform that enables organizations reduce the cost and complexity of managing quickly growing storage requirements. The Mezeo Cloud Storage Platform is a massively scalable solution that transforms traditional distributed file systems and object stores with their associated storage devices into a multi-tenant storage cloud. Mezeo provides for multiple storage technologies to seamlessly integrate into a single storage cloud, and provides solutions for cloud federation.

Year Founded	2009		Moody's Rating	N/A		Pricing	N

IaaS Sub-Classification

White Label
Traditional
Servers
Storage
Disaster Recovery
Backup
Messaging
Content Delivery
Desktop-as-a-Service
Hosting
IT-as-a-Service
Management Software

Direct & Channel

Hosting.com, Iomart Hosting, LayeredTech, Mezolink, OpSource, Softlayer, Triple C, VISI

Clients

None Listed

Summary

Termination Notification (Days)	N/A
Dedicated Account Management	N/A
Live Customer Support	Y
Guaranteed Network Availability	N/A
Root/ Administrator Access	N
Portal Support	Y

Updates

Available
Q2 2011

CloudSourcing100.com

Regulation and Compliance

N/A

Programming Languages

N/A

Operating Systems Compatibility

N/A

 **Q1 2011**

SOLUTION TYPE	PROVIDER TYPE | CUSTOMER CONTROL	

www.microsoft.com/windowsazure

PROVIDER SUMMARY

The Windows Azure Platform is a Microsoft cloud platform offering that enables customers to deploy applications & data into the cloud. This Platform is classified as platform as a service & forms part of Microsoft's cloud computing strategy, along with their software as a service offering, Microsoft Online Services. The platform appliance consists of Windows Azure, SQL Azure & a Microsoft-specified configuration of network, storage & server hardware. The appliance is designed for service providers, large enterprises & governments & provides a proven cloud platform that delivers breakthrough datacenter efficiency through innovative power, cooling & automation technologies.

Year Founded	1975		Moody's Rating	Aa1		Pricing	Y

IaaS Sub-Classification

White Label	✔
Traditional	
Servers	
Storage	
Disaster Recovery	
Backup	
Messaging	
Content Delivery	✔
Desktop-as-a-Service	
Hosting	
IT-as-a-Service	
Management Software	

Direct & Channel

Multiple Partner Listings

Clients

Multiple Client Listings

Summary

Termination Notification (Days)	N/A
Dedicated Account Management	N/A
Live Customer Support	Y
Guaranteed Network Availability	Y
Root/ Administrator Access	Y
Portal Support	Y

Updates

Available Q2 2011

CloudSourcing100.com

Regulation and Compliance

Multiple Compliance Listings

Operating Systems Compatibility

Windows Server 2003, Windows Server 2008

Programming Languages

Java, Ruby

ALSBRIDGE

CLOUD SOURCING 100   Q1 2011

SOLUTION TYPE	PROVIDER TYPE	CUSTOMER CONTROL

MirrorImage
www.mirror-image.com

PROVIDER SUMMARY

Mirror Image® Internet is a global network for online content, application and transaction delivery, and provides content delivery, streaming media, Web computing and reporting solutions that offer customers a way to create more engaging Web experiences for users worldwide. With an enterprise-class platform combining an optimal mix of connectivity, processing power and storage, Mirror Image provides retail, advertising, media and government organizations the control they need to deliver the right content to the right customer at the right time.

Year Founded	1997

Moody's Rating	N/A

Pricing	N

IaaS Sub-Classification

White Label	
Traditional	
Servers	
Storage	
Disaster Recovery	
Backup	
Messaging	
Content Delivery	✓
Desktop-as-a-Service	
Hosting	
IT-as-a-Service	
Management Software	

Direct & Channel

None Listed

Summary

Termination Notification (Days)	N/A
Dedicated Account Management	N/A
Live Customer Support	N/A
Guaranteed Network Availability	N/A
Root/ Administrator Access	N
Portal Support	N/A

Clients

NOAA
Forbes
Petstore.com
Orvis
Ansari Xprize

Updates

Available
Q2 2011

CloudSourcing100.com

Regulation and Compliance

N/A

Operating Systems Compatibility

N/A

Programming Languages

N/A

 C I N G 1 0 0   **Q1 2011**

SOLUTION TYPE

PROVIDER TYPE | CUSTOMER CONTROL

NaviSite®

www.navisite.com

PROVIDER SUMMARY

NaviSite, Inc. (acquired by Time Warner Cable February 2011) provides hosting, application management and managed cloud services for enterprises. They provide a full suite of reliable and scalable managed services, including Application Services, industry-leading Enterprise Hosting, and Managed Cloud Services. NaviSite's Managed Cloud Services (MCS) provides enterprise on-demand scalable provisioning of IT services including applications, messaging and collaboration, servers, storage, and networks.

Year Founded	1996

Moody's Rating	Baa2

Pricing	N

IaaS Sub-Classification

White Label	
Traditional	
Servers	✓
Storage	
Disaster Recovery	
Backup	
Messaging	✓
Content Delivery	
Desktop-as-a-Service	
Hosting	✓
IT-as-a-Service	
Management Software	

Direct & Channel

Cisco, IBM, Citrix, HP, Microsoft, Oracle, VMWare, Blackberry

Clients

None Listed

Summary

Termination Notification (Days)	N/A
Dedicated Account Management	N/A
Live Customer Support	Y
Guaranteed Network Availability	Y
Root/ Administrator Access	Y
Portal Support	Y

Updates

Available Q2 2011

CloudSourcing100.com

Regulation and Compliance

SAS70 Type II, PCI/DSS, HIPPA and more

Operating Systems Compatibility

N/A

Programming Languages

N/A

 CLOUD SOURCING 100 **Q1 2011**

SOLUTION TYPE	PROVIDER TYPE | CUSTOMER CONTROL	
		Nephoscale www.nephoscale.com

PROVIDER SUMMARY

NephoScale offers an integrated combination of software and hardware that provides users with control over a cloud infrastructure. By incorporating NephoScale's cloud services into applications, users can develop and deliver software faster, cheaper, and more dependably. Using CloudScript, a single uniform interface designed to dynamically provision and manipulate elements within a cloud infrastructure, user applications can create complex computing environments with a single API submission.

Year Founded	2009		Moody's Rating	N/A		Pricing	Y

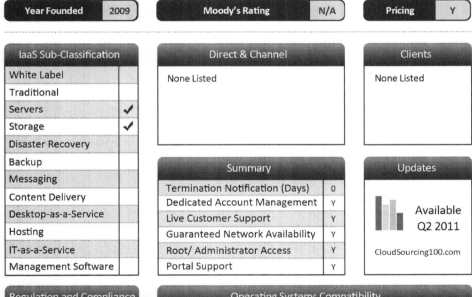

IaaS Sub-Classification	
White Label	
Traditional	
Servers	✔
Storage	✔
Disaster Recovery	
Backup	
Messaging	
Content Delivery	
Desktop-as-a-Service	
Hosting	
IT-as-a-Service	
Management Software	

Direct & Channel
None Listed

Clients
None Listed

Summary	
Termination Notification (Days)	0
Dedicated Account Management	Y
Live Customer Support	Y
Guaranteed Network Availability	Y
Root/ Administrator Access	Y
Portal Support	Y

Updates
Available Q2 2011
CloudSourcing100.com

Regulation and Compliance
SAS70 Type II

Operating Systems Compatibility
CentOS 5.x, Ubuntu Server 10.04, Windows Server 2008

Programming Languages
CloudScript

 C L O U D S O U R C I N G 100 **Q1 2011**

SOLUTION TYPE	PROVIDER TYPE | CUSTOMER CONTROL	
		 www.onlinetech.com

PROVIDER SUMMARY

Online Tech offers hosting solutions including basic colocation, managed colocation and managed dedicated servers. They deliver a range of data center services from basic colocation to managed servers from SAS 70-audited data centers. Online Tech's private cloud computing services combine the flexibility and cost-savings of cloud computing with the security, data integrity, and service level agreements (SLAs) of their SAS 70-certified environment.

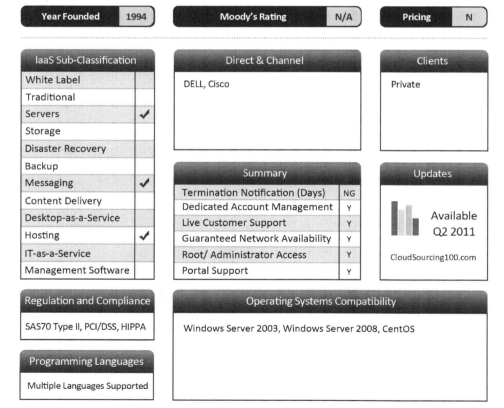

Year Founded	1994	Moody's Rating	N/A	Pricing	N

IaaS Sub-Classification	
White Label	
Traditional	
Servers	✔
Storage	
Disaster Recovery	
Backup	
Messaging	✔
Content Delivery	
Desktop-as-a-Service	
Hosting	✔
IT-as-a-Service	
Management Software	

Direct & Channel

DELL, Cisco

Clients

Private

Summary

Termination Notification (Days)	NG
Dedicated Account Management	Y
Live Customer Support	Y
Guaranteed Network Availability	Y
Root/ Administrator Access	Y
Portal Support	Y

Updates

Available
Q2 2011

CloudSourcing100.com

Regulation and Compliance

SAS70 Type II, PCI/DSS, HIPPA

Operating Systems Compatibility

Windows Server 2003, Windows Server 2008, CentOS

Programming Languages

Multiple Languages Supported

ALSBRIDGE

CLOUD SOURCING 100   Q1 2011

SOLUTION TYPE

PROVIDER TYPE | CUSTOMER CONTROL

 OpSource™
www.opsource.net

PROVIDER SUMMARY

OpSource™ provides cloud and managed hosting solutions for businesses of all sizes to accelerate growth and scale operations, while controlling costs and reducing IT infrastructure support risks. More than 400 Software-as-a-Service ISVs, cloud platform providers, carriers and enterprises rely on OpSource's expertise, experience and agility to operate high-availability, business-critical hosting environments. Their industry-leading Application Operations service goes beyond traditional hosting by providing application management, change management, performance management and application optimization.

Year Founded	2002		Moody's Rating	N/A		Pricing	Y

IaaS Sub-Classification

White Label	
Traditional	
Servers	✓
Storage	✓
Disaster Recovery	
Backup	
Messaging	
Content Delivery	
Desktop-as-a-Service	
Hosting	✓
IT-as-a-Service	
Management Software	

Direct & Channel

Akami, CentOS, Cisco, DELL, EMC, EQUI-NIX, Gomez, Hewlett-Packard, Microsoft, MySQL, NTT Group, Oracle, RedHat, SalesForce.com, VMWare

Summary

Termination Notification (Days)	3
Dedicated Account Management	Y
Live Customer Support	Y
Guaranteed Network Availability	Y
Root/ Administrator Access	Y
Portal Support	Y

Clients

Private

Updates

Available
Q2 2011

CloudSourcing100.com

Regulation and Compliance

SAS70 Type II, PCI DSS Level 1, European Safe Harbor, HIPAA

Programming Languages

Perl, PHP, Python, SQL

Operating Systems Compatibility

Cent OS, Cent OS 5.1, Cloud Optimised Cent OS, Linux Operating Systems, Red Hat 5.1, Red Hat Enterprise Linux, Ubuntu Linux, Windows Server 2003, Windows Server 2008

ALSBRIDGE

 **Q1 2011**

SOLUTION TYPE

PROVIDER TYPE | CUSTOMER CONTROL

Oracle
www.oracle.com

PROVIDER SUMMARY

Oracle's portfolio of cloud offerings includes Oracle On Demand, which provides software as a service, as well as hosted and managed alternatives to on-premise deployment. For enterprises that are building private clouds and for service providers that are building public clouds, Oracle offers comprehensive solutions for platform as a service and infrastructure as a service.

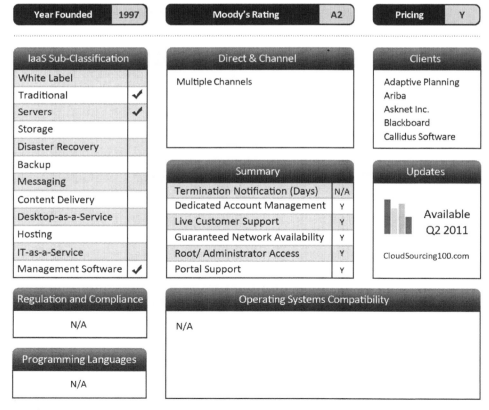

Year Founded	1997
Moody's Rating	A2
Pricing	Y

IaaS Sub-Classification
White Label	
Traditional	✔
Servers	✔
Storage	
Disaster Recovery	
Backup	
Messaging	
Content Delivery	
Desktop-as-a-Service	
Hosting	
IT-as-a-Service	
Management Software	✔

Direct & Channel
Multiple Channels

Clients
Adaptive Planning
Ariba
Asknet Inc.
Blackboard
Callidus Software

Summary
Termination Notification (Days)	N/A
Dedicated Account Management	Y
Live Customer Support	Y
Guaranteed Network Availability	Y
Root/ Administrator Access	Y
Portal Support	Y

Updates
Available Q2 2011
CloudSourcing100.com

Regulation and Compliance
N/A

Operating Systems Compatibility
N/A

Programming Languages
N/A

 **Q1 2011**

SOLUTION TYPE

PROVIDER TYPE | CUSTOMER CONTROL

OrangeScape
www.orangescape.com

PROVIDER SUMMARY

OrangeScape provides Platform as a Service (PaaS) to build domain solutions. OrangeScape uses a modeling driven visual development environment for creating business applications and can be deployed as SaaS or on-premise applications. The application deployed as SaaS runs on OrangeScape Cloud. The product is rewritten to run its development environment on the browser and to support cloud-oriented datastores such as BigTable. OrangeScape Cloud, leverages the cloud infrastructure of Google App Engine or Microsoft Azure (planned release) to provide massive scalability and storage capacity.

Year Founded	2003

Moody's Rating	N/A

Pricing	Y

IaaS Sub-Classification

White Label	
Traditional	
Servers	
Storage	
Disaster Recovery	
Backup	
Messaging	
Content Delivery	
Desktop-as-a-Service	
Hosting	
IT-as-a-Service	
Management Software	

Direct & Channel

Wipro, 3iInfotech, MphasiS, L&T Infotech

Summary

Termination Notification (Days)	N/A
Dedicated Account Management	N/A
Live Customer Support	Y
Guaranteed Network Availability	N/A
Root/ Administrator Access	Y
Portal Support	Y

Clients

Ford
Sterlite
Citi
TAIN
BIGRED Tubulars

Updates

 Available
Q2 2011

CloudSourcing100.com

Regulation and Compliance

N/A

Operating Systems Compatibility

N/A

Programming Languages

Java or Microsoft environments

ALSBRIDGE

 CING 100 **Q1 2011**

SOLUTION TYPE	PROVIDER TYPE	CUSTOMER CONTROL

Oxygen Cloud
www.oxygencloud.com

PROVIDER SUMMARY

Oxygen Cloud connects all people, data and devices to a single file system. They provide native desktop collaboration and cloud storage brokering to business end users. The company is a subsidiary of LeapFILE - the self-funded and profitable provider of secure file transfer to the Fortune 500 and companies of every size. They provide its users with direct desktop access to browse files in the cloud locally. With an intuitive desktop client, users can mix and match various public or private storage clouds for different business needs.

Year Founded	2010

Moody's Rating	N/A

Pricing	N

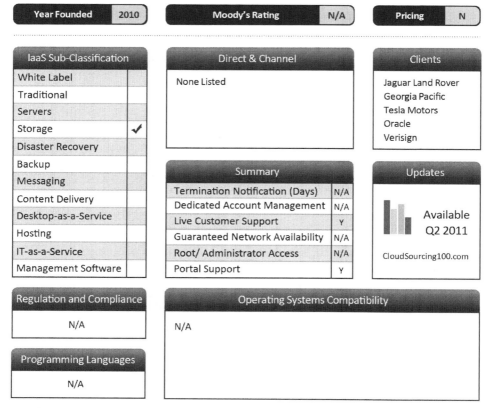

IaaS Sub-Classification	
White Label	
Traditional	
Servers	
Storage	✓
Disaster Recovery	
Backup	
Messaging	
Content Delivery	
Desktop-as-a-Service	
Hosting	
IT-as-a-Service	
Management Software	

Direct & Channel
None Listed

Clients
Jaguar Land Rover
Georgia Pacific
Tesla Motors
Oracle
Verisign

Summary	
Termination Notification (Days)	N/A
Dedicated Account Management	N/A
Live Customer Support	Y
Guaranteed Network Availability	N/A
Root/ Administrator Access	N/A
Portal Support	Y

Updates
Available Q2 2011
CloudSourcing100.com

Regulation and Compliance
N/A

Operating Systems Compatibility
N/A

Programming Languages
N/A

ALSBRIDGE

CLOUD SOURCING 100

Q1 2011

SOLUTION TYPE

   B

PROVIDER TYPE | CUSTOMER CONTROL

www.parallels.com

PROVIDER SUMMARY

Parallels provides virtualization and automation software for all major hardware, operating system, and virtualization platforms. Their virtualization strategy takes advantage of innovative hypervisor and container (Virtuozzo) technologies to serve the broadest range of scenarios and benefits possible. They also provide software providing hardware virtualization for Macintosh computers with Intel processors. The company offers a wide range of virtualization and automation solutions to help individuals and organizations of all sizes realize the benefits of optimized computing.

Year Founded	1999

Moody's Rating	N/A

Pricing	N

IaaS Sub-Classification

White Label	
Traditional	
Servers	✓
Storage	
Disaster Recovery	
Backup	
Messaging	
Content Delivery	
Desktop-as-a-Service	✓
Hosting	
IT-as-a-Service	
Management Software	

Direct & Channel

AMD, Apple, Savvis, Carinet, Codero and many more

Summary

Termination Notification (Days)	N/A
Dedicated Account Management	N/A
Live Customer Support	Y
Guaranteed Network Availability	Y
Root/ Administrator Access	Y
Portal Support	Y

Clients

None Listed

Updates

Available Q2 2011

CloudSourcing100.com

Regulation and Compliance

N/A

Programming Languages

C#, Java, Perl, Python

Operating Systems Compatibility

Cent OS, Fedora, Fedora 10, Fedora 11, FreeBSD, Gentoo 2008.0, Linux Operating Systems, SUSE Linux, Ubuntu Linux, Windows Server 2003, Windows Server 2008

ALSBRIDGE

CLOUD SOURCING 100 — Q1 2011

SOLUTION TYPE

PROVIDER TYPE | CUSTOMER CONTROL

PayPerCloud
www.paypercloud.com

PROVIDER SUMMARY

PayPerCloud's managed solutions are delivered to give secured and dynamic managed facilities for all hosting needs. Through managed collocation service, minimize the probable risks for clients server and can plan internet strategy well before any disaster occurs. PayPerCloud enables partners to offer a hosted services for small to mid-sized business customers. The key to reaching SMB's across the world is to arm partners to sell the complete service, under their name & brand.

Year Founded	2002

Moody's Rating	N/A

Pricing	Y

IaaS Sub-Classification

White Label	✓
Traditional	
Servers	✓
Storage	
Disaster Recovery	
Backup	
Messaging	✓
Content Delivery	
Desktop-as-a-Service	
Hosting	✓
IT-as-a-Service	
Management Software	

Direct & Channel

None Listed

Summary

Termination Notification (Days)	N/A
Dedicated Account Management	Y
Live Customer Support	Y
Guaranteed Network Availability	Y
Root/ Administrator Access	Y
Portal Support	Y

Clients

SAP
Xactly
Adobe
Ribbit
bmcSoftware

Updates

Available
Q2 2011

CloudSourcing100.com

Regulation and Compliance

SAS70 Type II, HIPPA

Operating Systems Compatibility

N/A

Programming Languages

N/A

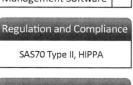

CLOUD SOURCING 100   **Q1 2011**

SOLUTION TYPE	PROVIDER TYPE | CUSTOMER CONTROL	
		 www.peer1.com

PROVIDER SUMMARY

Peer 1 is a hosting service provider that offers a broad range of internet infrastructure solutions including managed hosting, dedicated servers, colocation, and cloud computing to businesses all over the world. Peer 1 Hosting built out a 41,000-square-foot (3,800 m2) green data center in the Toronto area. This is thier most eco-friendly data center to date, implementing the most efficient products and technologies on the market. Their services are delivered over their SuperNetwork™ and provide highly scalable solutions.

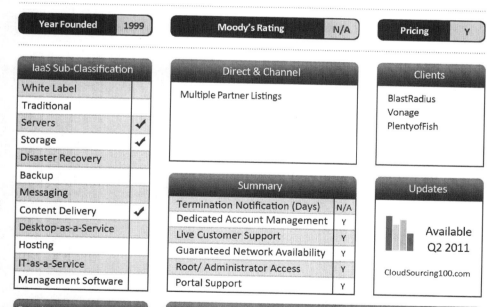

Year Founded	1999		Moody's Rating	N/A		Pricing	Y

IaaS Sub-Classification

White Label	
Traditional	
Servers	✔
Storage	✔
Disaster Recovery	
Backup	
Messaging	
Content Delivery	✔
Desktop-as-a-Service	
Hosting	
IT-as-a-Service	
Management Software	

Direct & Channel

Multiple Partner Listings

Clients

BlastRadius
Vonage
PlentyofFish

Summary

Termination Notification (Days)	N/A
Dedicated Account Management	Y
Live Customer Support	Y
Guaranteed Network Availability	Y
Root/ Administrator Access	Y
Portal Support	Y

Updates

Available
Q2 2011

CloudSourcing100.com

Regulation and Compliance

SAS70 Type II, CICA 5970,
Safe Harbor and more

Operating Systems Compatibility

Microsoft Windows, Red Hat Enterprise Linux

Programming Languages

Multiple Languages Supported

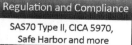

CLOUD SOURCING 100 **Q1 2011**

SOLUTION TYPE   PROVIDER TYPE | CUSTOMER CONTROL

            Platform Computing
www.platform.com

PROVIDER SUMMARY

Platform Computing offers solutions that address the unique, sophisticated and demanding requirements of High Performance Computing (HPC) environments in industry and research. The solutions are designed to help organizations increase productivity while minimizing the need for additional investment in hardware and infrastructure. Rapidly transform your internal IT data centers into dynamic private clouds (IaaS) from a single solution without lock-in.

Year Founded	1992

Moody's Rating	N/A

Pricing	N

IaaS Sub-Classification

White Label	
Traditional	
Servers	
Storage	
Disaster Recovery	
Backup	
Messaging	
Content Delivery	
Desktop-as-a-Service	
Hosting	
IT-as-a-Service	
Management Software	✓

Direct & Channel

Accelicon, Accerlrys, CSC, DELL, HP, Microsoft, RedHat and more

Clients

ARM
Pratt & Whitney
Infineon
Citigroup
Gaselys

Summary

Termination Notification (Days)	N/A
Dedicated Account Management	N/A
Live Customer Support	Y
Guaranteed Network Availability	N/A
Root/ Administrator Access	N
Portal Support	Y

Updates

Available
Q2 2011

CloudSourcing100.com

Regulation and Compliance

N/A

Operating Systems Compatibility

N/A

Programming Languages

N/A

CLOUD SOURCING 100   Q1 2011

SOLUTION TYPE	PROVIDER TYPE | CUSTOMER CONTROL	
		 www.rackspace.com

PROVIDER SUMMARY

Rackspace provides enterprise-level hosting services to businesses of all sizes and kinds around the world. Rackspace got started in 1998 and since have grown to serve more than 130,000 customers, including over 110,000 cloud computing customers. Rackspace integrates the technologies for each customer's need and delivers it as a service via the company's commitment to "Fanatical Support". Rackspace core products include Managed Hosting, Cloud Hosting and Email & Apps.

Year Founded	1998		Moody's Rating	N/A		Pricing	Y

IaaS Sub-Classification

White Label	
Traditional	
Servers	✔
Storage	✔
Disaster Recovery	
Backup	
Messaging	✔
Content Delivery	
Desktop-as-a-Service	
Hosting	✔
IT-as-a-Service	
Management Software	

Direct & Channel

Accenture, Citrix, Datalink, Magento, BMC Group, Siwel, Avanara, Gogoit and many more

Clients

Radio Flyer
Carlsberg
Plixi
Teen Choice 09
FreshBooks

Summary

Termination Notification (Days)	30
Dedicated Account Management	Y
Live Customer Support	Y
Guaranteed Network Availability	Y
Root/ Administrator Access	Y
Portal Support	Y

Updates

Available Q2 2011

CloudSourcing100.com

Regulation and Compliance

SAS70 Type II

Programming Languages

Multiple Languages Supported

Operating Systems Compatibility

Arch 2010.05, CentOS 5.5, CentOS 5.4, Debian 5.0, Fedora, Gentoo 10.1, Oracle EL R5U4, Oracle EL R5U3 JEOS, Red Hat EL 5.5, Red Hat EL 5.4, Ubuntu 10.10 LTS, Ubuntu 10.04 LTS, Ubuntu 9.10, Ubuntu 8.04.2 LTS

ALSBRIDGE

 **CLOUD** S O U R C I N G 100 **Q1 2011**

SOLUTION TYPE	PROVIDER TYPE | CUSTOMER CONTROL	
		 www.redhat.com

PROVIDER SUMMARY

Red Hat delivers infrastructure for cloud computing. Red Hat recognize that the IT infrastructure of most companies is - and will continue to be - composed of pieces from many different hardware and software vendors. Red Hat enables businesses to use and manage these diverse assets as one cloud. Red Hat enables cloud to be an evolution, not a revolution or a monolithic stack locked to the technology roadmap and business practices of a single vendor.

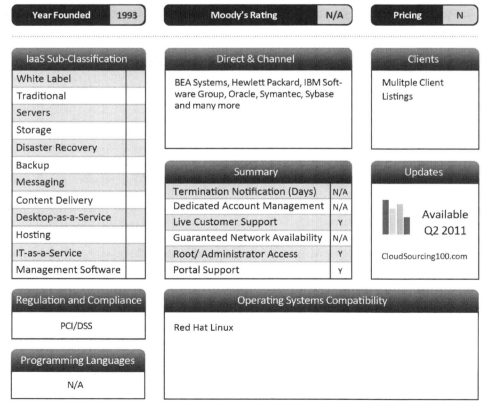

Year Founded	1993	Moody's Rating	N/A	Pricing	N

IaaS Sub-Classification

- White Label
- Traditional
- Servers
- Storage
- Disaster Recovery
- Backup
- Messaging
- Content Delivery
- Desktop-as-a-Service
- Hosting
- IT-as-a-Service
- Management Software

Direct & Channel

BEA Systems, Hewlett Packard, IBM Software Group, Oracle, Symantec, Sybase and many more

Clients

Mulitple Client Listings

Summary

Termination Notification (Days)	N/A
Dedicated Account Management	N/A
Live Customer Support	Y
Guaranteed Network Availability	N/A
Root/ Administrator Access	Y
Portal Support	Y

Updates

Available Q2 2011

CloudSourcing100.com

Regulation and Compliance

PCI/DSS

Operating Systems Compatibility

Red Hat Linux

Programming Languages

N/A

CLOUD SOURCING 100 Q1 2011

SOLUTION TYPE	PROVIDER TYPE | CUSTOMER CONTROL	
I P S B	F	H

Reliacloud
www.reliacloud.com

PROVIDER SUMMARY

ReliaCloud is a cloud infrastructure service that hosts cloud servers and cloud storage in a variety of Windows and Linux environments. ReliaCloud features an API, load balancing, a hardware firewall, and high availability. They serve more than 10,000 customers with products and services through two SAS 70 Type II data centers in St. Paul and Eden Prairie, Minnesota. The company has long provided a full range of Web and application hosting services with service and support, and in 2009 introduced its new suite of cloud computing services under the ReliaCloud brand.

Year Founded	2009		Moody's Rating	N/A		Pricing	Y

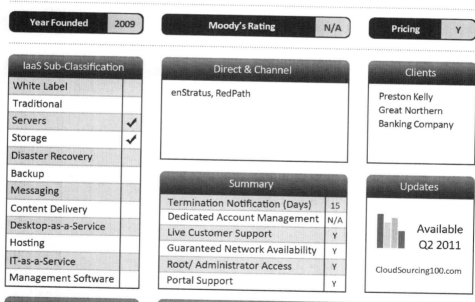

IaaS Sub-Classification	
White Label	
Traditional	
Servers	✓
Storage	✓
Disaster Recovery	
Backup	
Messaging	
Content Delivery	
Desktop-as-a-Service	
Hosting	
IT-as-a-Service	
Management Software	

Direct & Channel
enStratus, RedPath

Clients
Preston Kelly
Great Northern
Banking Company

Summary	
Termination Notification (Days)	15
Dedicated Account Management	N/A
Live Customer Support	Y
Guaranteed Network Availability	Y
Root/ Administrator Access	Y
Portal Support	Y

Updates
Available Q2 2011
CloudSourcing100.com

Regulation and Compliance
SAS70 Type II

Operating Systems Compatibility
CentOS, Debian, FreeBSD, Ubuntu, Windows Server 2003, Windows Server 2008

Programming Languages
Multiple Languages Supported

ALSBRIDGE

 **Q1 2011**

SOLUTION TYPE	PROVIDER TYPE | CUSTOMER CONTROL	

Cloud Management Platform
www.rightscale.com

PROVIDER SUMMARY

RightScale is a web based cloud computing management platform for managing cloud infrastructure from multiple providers. The company offers a fully automated management platform that delivers the scalable, cost-effective, on-demand power of cloud computing, while providing complete IT control and transparency. The web-based RightScale Cloud Management Platform is available in a range of editions and solution packs. Thousands of deployments and over 2,000,000 servers have been launched on RightScale for leading companies such as Animoto, Playfish, Sling Media and TC3.

Year Founded	2006

Moody's Rating	N/A

Pricing	N

IaaS Sub-Classification

White Label	
Traditional	
Servers	
Storage	
Disaster Recovery	
Backup	
Messaging	
Content Delivery	
Desktop-as-a-Service	
Hosting	
IT-as-a-Service	
Management Software	✔

Direct & Channel

IBM, Jaspersoft, MySQL, Capgemini, Hitachi Solutions, HyperStratus, ProKarma, Amazon Web Services and many more

Summary

Termination Notification (Days)	N/A
Dedicated Account Management	N/A
Live Customer Support	Y
Guaranteed Network Availability	N/A
Root/ Administrator Access	Y
Portal Support	Y

Clients

Zynga
A&E
Fansnap
PBS
AmericanGirl

Updates

Available
Q2 2011

CloudSourcing100.com

Regulation and Compliance

N/A

Operating Systems Compatibility

Ubuntu, CentOS, Windows 2003, Windows 2008

Programming Languages

N/A

CLOUD SOURCING 100 — Q1 2011

SOLUTION TYPE

PROVIDER TYPE | CUSTOMER CONTROL

rPath
www.rpath.com

PROVIDER SUMMARY

rPath, Inc. is a technology company based in Raleigh, North Carolina that provides a platform for enterprise IT organizations, independent software vendors (ISVs) and on-demand service providers to automate the process of constructing (or packaging), deploying and updating software stacks across physical, virtual and cloud-based environments. rPath provides a low-overhead solution for consistency and control in software systems construction, deployment and change across physical, virtual and cloud-based environments.

Year Founded	1982	Moody's Rating	N/A	Pricing	N

IaaS Sub-Classification

White Label	
Traditional	
Servers	
Storage	
Disaster Recovery	
Backup	
Messaging	
Content Delivery	
Desktop-as-a-Service	
Hosting	
IT-as-a-Service	
Management Software	✔

Direct & Channel

Eucalypts, NewScale, RightScale, Rackspace, VMWare, Collabnet, Amazon Web Services, Bluelock, Novell, Citrix, OpSource

Summary

Termination Notification (Days)	N/A
Dedicated Account Management	N/A
Live Customer Support	N/A
Guaranteed Network Availability	N/A
Root/ Administrator Access	Y
Portal Support	Y

Clients

AMD
Imageworks
Qualcomm
Fujitsu
IBM

Updates

Available
Q2 2011

CloudSourcing100.com

Regulation and Compliance

N/A

Programming Languages

N/A

Operating Systems Compatibility

Red Hat Enterprise Linux (4, 5), Novell SUSE Linux (10, 11), CentOS (5), Windows Server 2003, Windows Server 2008

CLOUD SOURCING 100

Q1 2011

SOLUTION TYPE	PROVIDER TYPE | CUSTOMER CONTROL	
I **P** **S** B	**F** **L**	The Security Division of EMC www.rsa.com

PROVIDER SUMMARY

The RSA Security Practice of EMC Consulting offers broad based security assessments for virtualized environments and new services to secure Virtual Desktop Infrastructures, leverage RSA best practices and established safeguards to help build secure virtualized and private cloud environments through technology, policy and program development. Their services apply a range of authentication, data protection, and security event management solutions into virtualized environments to accelerate the return on investment of this technology while maintaining the appropriate security controls and posture.

Year Founded	2006

Moody's Rating	N/A

Pricing	N

IaaS Sub-Classification

White Label	
Traditional	✓
Servers	
Storage	
Disaster Recovery	
Backup	
Messaging	
Content Delivery	
Desktop-as-a-Service	
Hosting	
IT-as-a-Service	
Management Software	✓

Direct & Channel

Multiple Partner Listings

Clients

Mulitple Client Listings

Summary

Termination Notification (Days)	NG
Dedicated Account Management	Y
Live Customer Support	Y
Guaranteed Network Availability	N/A
Root/ Administrator Access	N
Portal Support	Y

Updates

Available
Q2 2011

CloudSourcing100.com

Regulation and Compliance

Multiple Compliance Listings

Operating Systems Compatibility

Red Hat Enterprise Linux (4, 5), Sun Solaris, Windows Server 2003

Programming Languages

N/A

ALSBRIDGE

CLOUD SOURCING 100 — Q1 2011

SOLUTION TYPE

   I P S B

PROVIDER TYPE | CUSTOMER CONTROL

 F L

www.salesforce.com

PROVIDER SUMMARY

Salesforce.com an enterprise cloud computing company distributes business software on a subscription basis. Salesforce.com hosts the applications offsite. It is best known for its Customer Relationship Management (CRM) products and, through acquisition, has expanded into the "social enterprise arena". Salesforce.com's CRM solution is broken down into several broad categories: Sales Cloud, Service Cloud, Data Cloud(including Jigsaw), Collaboration Cloud (including Chatter) and Custom Cloud (including Force.com).

Year Founded	1999	Moody's Rating	N/A	Pricing	N

IaaS Sub-Classification

White Label
Traditional
Servers
Storage
Disaster Recovery
Backup
Messaging
Content Delivery
Desktop-as-a-Service
Hosting
IT-as-a-Service
Management Software

Direct & Channel

3000+ successful partners and counting

Summary

Termination Notification (Days)	N/A
Dedicated Account Management	Y
Live Customer Support	Y
Guaranteed Network Availability	Y
Root/ Administrator Access	N
Portal Support	Y

Clients

Mulitple Client Listings

Updates

Available Q2 2011

CloudSourcing100.com

Regulation and Compliance

N/A

Programming Languages

N/A

Operating Systems Compatibility

N/A

ALSBRIDGE

 **CLOUD** S O U R C I N G 100 **Q1 2011**

SOLUTION TYPE	PROVIDER TYPE | CUSTOMER CONTROL	

www.savvisknowscloud.com

PROVIDER SUMMARY

Savvis Symphony Virtual Private Data Center (VPDC) provides one of the industry's first enter-prise-class (VPDC) solutions with multi-tiered security and service profiles. A VPDC can contain a complete set of enterprise data center services, including compute instances of varying sizes, multiple tiers of storage, a wide variety of security features, high-performance, redundant band-width and load balancing.

Year Founded	1998		Moody's Rating	B1		Pricing	N

IaaS Sub-Classification

White Label	
Traditional	✔
Servers	✔
Storage	
Disaster Recovery	
Backup	
Messaging	
Content Delivery	
Desktop-as-a-Service	
Hosting	
IT-as-a-Service	
Management Software	

Direct & Channel

None Listed

Clients

Innovest
Hallmark
Discovery
Wallstreet

Summary

Termination Notification (Days)	NG
Dedicated Account Management	Y
Live Customer Support	Y
Guaranteed Network Availability	Y
Root/ Administrator Access	N/A
Portal Support	Y

Updates

Available
Q2 2011

CloudSourcing100.com

Regulation and Compliance

SAS70 Type II, PCI/DSS

Operating Systems Compatibility

Red Hat Linux 5, Microsoft Windows 2008

Programming Languages

N/A

ALSBRIDGE

CLOUD SOURCING 100

Q1 2011

SOLUTION TYPE

PROVIDER TYPE | CUSTOMER CONTROL

ServerCentral
www.servercentral.com

PROVIDER SUMMARY

ServerCentral is a managed data center solutions provider specializing in bandwidth-intensive colocation applications. With data centers located worldwide, ServerCentral provides services to a diverse roster of clients, including content delivery companies, web hosting and application service providers, hardware vendors and enterprise-level clients. SeverCentral focuses on providing services to their clients, no matter the size or scope of their company. Whether hosting a dedicated server package, distributing gigabits of media, or overseeing a large-scale colocation operation.

Year Founded	2000		Moody's Rating	N/A		Pricing	N

IaaS Sub-Classification

White Label	
Traditional	
Servers	✓
Storage	
Disaster Recovery	
Backup	
Messaging	
Content Delivery	
Desktop-as-a-Service	
Hosting	✓
IT-as-a-Service	
Management Software	

Direct & Channel

None Listed

Summary

Termination Notification (Days)	N/A
Dedicated Account Management	N
Live Customer Support	Y
Guaranteed Network Availability	N/A
Root/ Administrator Access	N/A
Portal Support	Y

Clients

StreamGuys
Ars Technica
CAT
Namco
SHCDIRECT

Updates

Available Q2 2011

CloudSourcing100.com

Regulation and Compliance

SAS70 Type II

Programming Languages

None Listed

Operating Systems Compatibility

N/A

ALSBRIDGE

CLOUD SOURCING 100   Q1 2011

SOLUTION TYPE	PROVIDER TYPE | CUSTOMER CONTROL	

sgi
www.sgi.com

PROVIDER SUMMARY

Cyclone™, SGI's cloud computing offering for technical applications capitalizes on over twenty years of SGI High Performance Computing (HPC) expertise. It specifically addresses the growing science and engineering markets that rely on high-end computational hardware, software and networking equipment to achieve rapid results. SGI systems run on some of the world's fastest supercomputing hardware architectures with robust storage for integrated scratch space and long term data archiving.

Year Founded	1980

Moody's Rating	N/A

Pricing	N

IaaS Sub-Classification

White Label	
Traditional	
Servers	
Storage	
Disaster Recovery	
Backup	
Messaging	
Content Delivery	
Desktop-as-a-Service	
Hosting	
IT-as-a-Service	
Management Software	

Direct & Channel

Multiple Partner Listings

Clients

AOL
Facebook
Yahoo
YouTube
Western Union

Summary

Termination Notification (Days)	N/A
Dedicated Account Management	N/A
Live Customer Support	Y
Guaranteed Network Availability	N/A
Root/ Administrator Access	N/A
Portal Support	N/A

Updates

Available
Q2 2011

CloudSourcing100.com

Regulation and Compliance

N/A

Operating Systems Compatibility

N/A

Programming Languages

None Listed

CLOUD SOURCING 100   Q1 2011

SOLUTION TYPE	PROVIDER TYPE | CUSTOMER CONTROL	
		SIMtone

www.simtone.net

PROVIDER SUMMARY

SIMtone has developed and commercialized the patented SIMtone Universal Cloud Computing Platform that allows network operators and businesses to host, manage and provision cloud-hosted services, and ubiquitously deliver them to zero-touch terminals that can be standalone, low cost hardware appliances, or software terminals usable via browsers or on PCs, thin clients and mobile devices. The SIMtone SNAP zero-touch cloud computing terminals are non- processing, fully stateless devices that allow end users to login to all of their cloud computing services with a single user ID from anywhere.

Year Founded	2006		Moody's Rating	N/A		Pricing	N

IaaS Sub-Classification

White Label	
Traditional	
Servers	
Storage	
Disaster Recovery	
Backup	
Messaging	
Content Delivery	✔
Desktop-as-a-Service	
Hosting	
IT-as-a-Service	
Management Software	

Direct & Channel

None Listed

Clients

None Listed

Summary

Termination Notification (Days)	N/A
Dedicated Account Management	N/A
Live Customer Support	N/A
Guaranteed Network Availability	N/A
Root/ Administrator Access	N/A
Portal Support	Y

Updates

Available
Q2 2011

CloudSourcing100.com

Regulation and Compliance

N/A

Operating Systems Compatibility

N/A

Programming Languages

None Listed

CLOUD SOURCING 100

Q1 2011

SOLUTION TYPE	PROVIDER TYPE	CUSTOMER CONTROL

www.softlayer.com

PROVIDER SUMMARY

SoftLayer provides on-demand web hosting and data center services. SoftLayer lets customers create dedicated, cloud, or seamless hybrid computing environments, leveraging world-class data centers in Dallas, Houston, Seattle, and Washington D.C., and network Points of Presence nationwide. SoftLayer automates all elements of its platform, empowering enterprises of any size with complete control, security, scalability, and ease-of-management via a leading Customer Portal and Open API.

Year Founded	2005		Moody's Rating	B2		Pricing	Y

IaaS Sub-Classification

White Label	
Traditional	
Servers	✔
Storage	✔
Disaster Recovery	
Backup	
Messaging	
Content Delivery	✔
Desktop-as-a-Service	
Hosting	
IT-as-a-Service	
Management Software	

Direct & Channel

Citrix Systems, Cloudkick, McAfee Inc., Microsoft, VeriSign, Red Hat, Parallels, SuperMicro, SendGrid and many more

Clients

ERICA Ltd.
MidPhase Services
MMO Guildsites
Pick-A-Prof
SlideShare

Summary

Termination Notification (Days)	30
Dedicated Account Management	N/A
Live Customer Support	Y
Guaranteed Network Availability	Y
Root/ Administrator Access	Y
Portal Support	Y

Updates

Available
Q2 2011

CloudSourcing100.com

Regulation and Compliance

IPv6 Ready, SAS70 TYPE II, PCI/DSS

Programming Languages

N/A

Operating Systems Compatibility

CentOS, RedHat Enterprise Linux, Citrix XenApp, Debian Etch, Debian Sarge, Oracle Enterprise Linux, SUSE Linux Enterprise Server, Microsoft Windows 2000, Microsoft Windows Server 2003, Microsoft Windows Vista, Microsoft Windows XP

ALSBRIDGE

CLOUD SOURCING 100   Q1 2011

SOLUTION TYPE	PROVIDER TYPE | CUSTOMER CONTROL	
I P S B	F	H

www.stormondemand.com

PROVIDER SUMMARY

Storm On Demand is a wholly owned subsidiary of Liquid Web Inc, a managed hosting and data center company. Storm servers are deployed within Liquid Web's 90,000 square foot state-of-the-art Cloud Data Center in the mid-western United States. The Storm On Demand platform is a proprietary cloud computing and server hosting infrastructure developed by the engineers at Liquid Web Inc. Storm On Demand enables users to deploy and manage cloud servers from the browser-based Storm Dashboard. Storm servers are backed by the engineering and support services of the Liquid Web Support team which is on site and available 24x7x365.

Year Founded	1997		Moody's Rating	N/A		Pricing	Y

IaaS Sub-Classification

White Label	
Traditional	
Servers	✔
Storage	
Disaster Recovery	
Backup	
Messaging	
Content Delivery	
Desktop-as-a-Service	
Hosting	✔
IT-as-a-Service	
Management Software	

Direct & Channel

None Listed

Clients

None Listed

Summary

Termination Notification (Days)	N/A
Dedicated Account Management	N/A
Live Customer Support	Y
Guaranteed Network Availability	Y
Root/ Administrator Access	N/A
Portal Support	Y

Updates

Available
Q2 2011
CloudSourcing100.com

Regulation and Compliance

N/A

Operating Systems Compatibility

CentOS, Debian, Ubuntu

Programming Languages

N/A

ALSBRIDGE

 CLOUD S O U R C I N G 100 **Q1 2011**

SOLUTION TYPE	PROVIDER TYPE | CUSTOMER CONTROL	
		StrataScale www.stratascale.com

PROVIDER SUMMARY

StrataScale's shared cloud infrastructure delivers flexibility, scalability, and affordability. Perfect for development and test environments, web hosting, short-term projects. StrataScale provides users with anywhere, anytime web portal control of their VM's to scale their cloud infrastructure on-demand. Bundled packages with everything you need to be up and running in minutes, manage and customize on-demand in real-time via advanced UI portal, superior security and guaranteed uptime, with enterprise-class performance.

Year Founded	2008		Moody's Rating	N/A		Pricing	N

IaaS Sub-Classification

White Label	✔
Traditional	
Servers	✔
Storage	
Disaster Recovery	
Backup	
Messaging	
Content Delivery	
Desktop-as-a-Service	
Hosting	✔
IT-as-a-Service	
Management Software	

Direct & Channel

None Listed

Clients

Resources
Doctor Dispense
Wintesnsive Technologies

Summary

Termination Notification (Days)	N/A
Dedicated Account Management	N/A
Live Customer Support	Y
Guaranteed Network Availability	Y
Root/ Administrator Access	N/A
Portal Support	Y

Updates

Available
Q2 2011

CloudSourcing100.com

Regulation and Compliance

SAS70 Type II

Operating Systems Compatibility

CentOS, RedHat Enterprise Linux, Debian Microsoft Windows Server 2003, Microsoft Windows Server 2008

Programming Languages

N/A

ALSBRIDGE

CLOUD SOURCING 100   Q1 2011

SOLUTION TYPE	PROVIDER TYPE | CUSTOMER CONTROL		TCS
B			www.tcs.com

PROVIDER SUMMARY

Tata Consultancy (TCS) is a software services consulting company headquartered in Mumbai, India. TCS is the largest provider of information technology and business process outsourcing services in Asia.TCS offers end-to-end Cloud Computing services with strategic consulting to further transform the way organizations leverage IT. TCS uses a structured approach to establish their clients business needs and the underlying technology infrastructure and thne works with the client to deploy the appropriate cloud computing components to deliver reduced costs, enhanced agility, quick service delivery and improved scalability.

Year Founded	1968	Moody's Rating	N/A	Pricing	N

IaaS Sub-Classification
White Label
Traditional
Servers
Storage
Disaster Recovery
Backup
Messaging
Content Delivery
Desktop-as-a-Service
Hosting
IT-as-a-Service
Management Software

Direct & Channel
None Listed

Clients
None Listed

Summary
Termination Notification (Days)	N/A
Dedicated Account Management	Y
Live Customer Support	Y
Guaranteed Network Availability	N/A
Root/ Administrator Access	N/A
Portal Support	N/A

Updates
Available
Q2 2011

CloudSourcing100.com

Regulation and Compliance
N/A

Programming Languages
N/A

Operating Systems Compatibility
N/A

ALSBRIDGE

 C I N G 100 **Q1 2011**

SOLUTION TYPE	PROVIDER TYPE | CUSTOMER CONTROL	
		 www.thinkgrid.co.uk

PROVIDER SUMMARY

ThinkGrid is a private corporation headquartered in the UK, with offices located in London, Seattle, Sydney, Belgrade and Lviv. ThinkGrid is a channel focused vendor providing turnkey worldwide infrastructure and a control panel platform that allows IT solution providers to become cloud service providers; able to create, bill and manage IT services from the cloud. ThinkGrid operates tier 3 & 4 data centers across the world to house their platform, which delivers IT services such as Hosted Virtual Desktop, Virtual Server, Voice to their partners.

Year Founded	2008		Moody's Rating	N/A		Pricing	N

IaaS Sub-Classification

White Label	
Traditional	
Servers	✓
Storage	✓
Disaster Recovery	✓
Backup	
Messaging	✓
Content Delivery	
Desktop-as-a-Service	✓
Hosting	
IT-as-a-Service	
Management Software	

Direct & Channel

Multiple Partner Listings

Clients

Pensar
Heythrop College
Netstar
ILG
LOKAHI Foundation

Summary

Termination Notification (Days)	30
Dedicated Account Management	N/A
Live Customer Support	Y
Guaranteed Network Availability	Y
Root/ Administrator Access	Y
Portal Support	Y

Updates

Available
Q2 2011

CloudSourcing100.com

Regulation and Compliance

N/A

Programming Languages

None Listed

Operating Systems Compatibility

Multiple Operating Systems

CLOUD SOURCING 100 — Q1 2011

SOLUTION TYPE	PROVIDER TYPE | CUSTOMER CONTROL	
		 www.unisys.com

PROVIDER SUMMARY

Unisys Secure Cloud Solution is a service which supports customers with a speed to provisioning, elastic capacity, and security for data "in motion". Unisys Secure Cloud enables users to leverage the power and flexibility of cloud computing to provision software applications, infrastructure and platforms on an as-needed basis with subscription-based pricing options. Unisys Secure Cloud is built upon a patent-pending security technology foundation that breaks new ground in data protection. This allows Unisys Secure Cloud customers to run much more of their business application workloads than typical commodity cloud offerings.

Year Founded	1986	Moody's Rating	B1	Pricing	N

IaaS Sub-Classification

White Label	
Traditional	✓
Servers	✓
Storage	
Disaster Recovery	
Backup	
Messaging	
Content Delivery	
Desktop-as-a-Service	
Hosting	
IT-as-a-Service	
Management Software	

Direct & Channel

None Listed

Clients

None Specific to Cloud

Summary

Termination Notification (Days)	NG
Dedicated Account Management	N/A
Live Customer Support	N/A
Guaranteed Network Availability	N/A
Root/ Administrator Access	N/A
Portal Support	N/A

Updates

Available Q2 2011

CloudSourcing100.com

Regulation and Compliance

ISO 20000 & 27001-certified delivery

Operating Systems Compatibility

N/A

Programming Languages

None Listed

 **Q1 2011**

SOLUTION TYPE

PROVIDER TYPE | CUSTOMER CONTROL

Vaultwise
www.vaultwise.com

PROVIDER SUMMARY

Vaultwise provides premium server backup services with its Corporate Backup and SolidState Backup products. Vaultwise offers complete disaster recovery solutions. Information that includes data from application aware back up, like SQL server table level, to in-service production hardware with device block level operations can be stored using VaultWise that offers state of the art storage. Supporting clients across a wide variety of business they have developed a strong network of services for all customers. With high rankings in the Dun & Bradstreet index you be confident your data is safe.

Year Founded	N/A		Moody's Rating	N/A		Pricing	N

IaaS Sub-Classification

White Label	
Traditional	
Servers	
Storage	
Disaster Recovery	✔
Backup	✔
Messaging	
Content Delivery	
Desktop-as-a-Service	
Hosting	
IT-as-a-Service	
Management Software	

Direct & Channel

None Listed

Clients

None Listed

Summary

Termination Notification (Days)	0
Dedicated Account Management	N/A
Live Customer Support	Y
Guaranteed Network Availability	Y
Root/ Administrator Access	N/A
Portal Support	Y

Updates

Available
Q2 2011

CloudSourcing100.com

Regulation and Compliance

HIPAA, Sarbanes Oxley Act, GLB. and more

Programming Languages

None Listed

Operating Systems Compatibility

Windows 95/98/ME/NT/2000/XP/2003, Linux Kernel 2.2 or above (e.g. Redhat Linux 6.x or above), Solaris 2.x or above, Mac OS X, Netware 5.1 or above, and all other platforms supporting Java2 Runtime 1.3.1 or above.

CLOUD SOURCING 100   Q1 2011

SOLUTION TYPE	PROVIDER TYPE | CUSTOMER CONTROL	
		vCloudExpress www.vcloudexpress.terremark.com

PROVIDER SUMMARY

vCloud Express is brought to you by Terremark (now owned by Verizon as of January 2011) and VMware, two of the leading names in cloud computing, both with a long history of partnering to bring state-of-the-art solutions to leading companies, governmental agencies and other organizations around the world. vCloud Express features a simple Web-based console based on Terremark's class-leading Enteprise Cloud platform. A full-featured API allows programmatic access to compute capacity. Terremark is a leading global provider of IT infrastructure services, delivered on the industry's most robust and advanced technology platform.

Year Founded	N/A

Moody's Rating	B2

Pricing	Y

IaaS Sub-Classification

White Label	
Traditional	
Servers	✓
Storage	
Disaster Recovery	
Backup	
Messaging	
Content Delivery	
Desktop-as-a-Service	
Hosting	
IT-as-a-Service	
Management Software	

Direct & Channel

None Listed

Clients

None Listed

Summary

Termination Notification (Days)	N/A
Dedicated Account Management	N/A
Live Customer Support	Y
Guaranteed Network Availability	Y
Root/ Administrator Access	Y
Portal Support	Y

Updates

Available
Q2 2011

CloudSourcing100.com

Regulation and Compliance

N/A

Programming Languages

N/A

Operating Systems Compatibility

Windows, Red Hat, CentOS and Ubuntu—or use our Blank Server deployment system and install one of 450-plus operating systems/versions that are compatible.

ALSBRIDGE

© Alsbridge Inc, 2011

CLOUD SOURCING 100

Q1 2011

SOLUTION TYPE	PROVIDER TYPE	CUSTOMER CONTROL

Verizon
www.verizonbusiness.com

PROVIDER SUMMARY

Verizon cloud IT services, provides enterprise-class computing, storage, and backup services over their global IP network on an as-needed basis. Industry analyst firm Current Analysis awarded Verizon cloud security services its highest rating, citing the breadth, depth and global availability across the company's software, platform, and infrastructure as-a-service offerings.

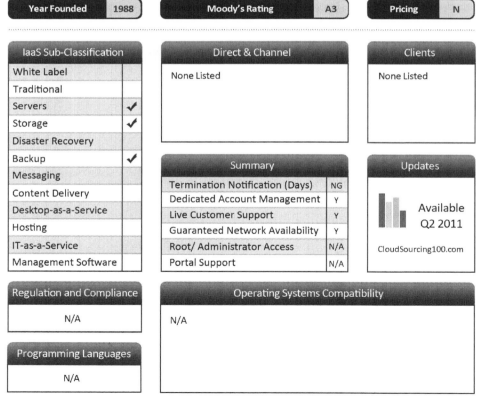

Year Founded	1988

Moody's Rating	A3

Pricing	N

IaaS Sub-Classification	
White Label	
Traditional	
Servers	✓
Storage	✓
Disaster Recovery	
Backup	✓
Messaging	
Content Delivery	
Desktop-as-a-Service	
Hosting	
IT-as-a-Service	
Management Software	

Direct & Channel
None Listed

Clients
None Listed

Summary	
Termination Notification (Days)	NG
Dedicated Account Management	Y
Live Customer Support	Y
Guaranteed Network Availability	Y
Root/ Administrator Access	N/A
Portal Support	N/A

Updates
Available Q2 2011
CloudSourcing100.com

Regulation and Compliance
N/A

Operating Systems Compatibility
N/A

Programming Languages
N/A

ALSBRIDGE

CLOUD SOURCING 100   Q1 2011

SOLUTION TYPE	PROVIDER TYPE | CUSTOMER CONTROL	

Vertica
www.vertica.com

PROVIDER SUMMARY

Vertica was designed specifically to scale and run 24×7 in grid-based computing environments like the cloud. It is the only cloud-based analytic database, which enable it to manage terabytes to petabytes of data faster with more reliably than any other cloud database. Vertica's goal was to revolutionize the relational database industry by developing a tool to solve today's analytical challenges. Vertica accomplished just that by building a new system from the ground up.

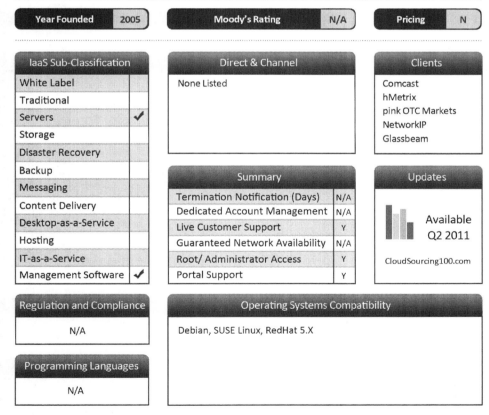

Year Founded	2005

Moody's Rating	N/A

Pricing	N

IaaS Sub-Classification
White Label	
Traditional	
Servers	✔
Storage	
Disaster Recovery	
Backup	
Messaging	
Content Delivery	
Desktop-as-a-Service	
Hosting	
IT-as-a-Service	
Management Software	✔

Direct & Channel
None Listed

Clients
Comcast
hMetrix
pink OTC Markets
NetworkIP
Glassbeam

Summary
Termination Notification (Days)	N/A
Dedicated Account Management	N/A
Live Customer Support	Y
Guaranteed Network Availability	N/A
Root/ Administrator Access	Y
Portal Support	Y

Updates
Available Q2 2011

CloudSourcing100.com

Regulation and Compliance
N/A

Operating Systems Compatibility
Debian, SUSE Linux, RedHat 5.X

Programming Languages
N/A

ALSBRIDGE

 CLOUD SOURCING 100 **Q1 2011**

SOLUTION TYPE	PROVIDER TYPE | CUSTOMER CONTROL	

www.virtela.net

PROVIDER SUMMARY

Virtela is a managed network, security and cloud services company providing services, including MPLS and IP Virtual Private Networks (VPNs), Security, Remote Infrastructure Management, and Application Acceleration, to midmarket and Fortune 500 customers globally. This unique model integrates best-of-breed technologies and the best local, regional and global networks, through their partnerships with over 500 carriers, to offer unparalleled geographic reach in over 190 countries. Virtela is the single point of contact for expert IT infrastructure design, implementation and 24x7 proactive monitoring and management worldwide.

Year Founded	2001

Moody's Rating	N/A

Pricing	N

IaaS Sub-Classification

White Label	
Traditional	
Servers	
Storage	
Disaster Recovery	
Backup	
Messaging	
Content Delivery	✔
Desktop-as-a-Service	
Hosting	
IT-as-a-Service	
Management Software	

Direct & Channel

None Listed

Clients

FedEx
Pitney Bowes
ING
Comcast
Google

Summary

Termination Notification (Days)	N/A
Dedicated Account Management	N/A
Live Customer Support	N/A
Guaranteed Network Availability	Y
Root/ Administrator Access	N/A
Portal Support	N/A

Updates

Available
Q2 2011

CloudSourcing100.com

Regulation and Compliance

N/A

Operating Systems Compatibility

N/A

Programming Languages

None Listed

CLOUD SOURCING 100   Q1 2011

SOLUTION TYPE	PROVIDER TYPE | CUSTOMER CONTROL	

www.vmware.com

PROVIDER SUMMARY

VMware the global leader in virtualization and cloud infrastructure, delivers customer-proven solutions that accelerate IT by reducing complexity and enabling more flexible, agile service delivery. VMware enables enterprises to adopt a cloud model that addresses their unique business challenges. VMware's approach accelerates the transition to cloud computing while preserving existing investments and improving security and control.

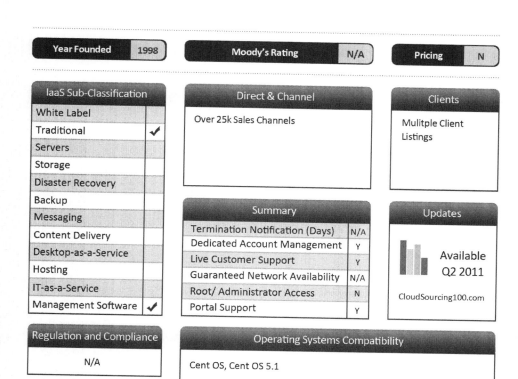

Year Founded	1998	Moody's Rating	N/A	Pricing	N

IaaS Sub-Classification

White Label	
Traditional	✔
Servers	
Storage	
Disaster Recovery	
Backup	
Messaging	
Content Delivery	
Desktop-as-a-Service	
Hosting	
IT-as-a-Service	
Management Software	✔

Direct & Channel

Over 25k Sales Channels

Clients

Mulitple Client Listings

Summary

Termination Notification (Days)	N/A
Dedicated Account Management	Y
Live Customer Support	Y
Guaranteed Network Availability	N/A
Root/ Administrator Access	N
Portal Support	Y

Updates

Available Q2 2011

CloudSourcing100.com

Regulation and Compliance

N/A

Operating Systems Compatibility

Cent OS, Cent OS 5.1

Programming Languages

Java, Perl, PHP, SQL

ALSBRIDGE

 **Q1 2011**

SOLUTION TYPE	PROVIDER TYPE	CUSTOMER CONTROL

www.voxel.net

PROVIDER SUMMARY

VoxCLOUD is the only cloud that unifies the physical and virtual across 3 continents, all supported 24x7x365 by Voxel expert support. VoxCLOUD is global, hybrid cloud computing offering. It enables customers to scale their applications up and down themselves. Voxel own and operate an International network connecting nearly 20 Voxel Network POPs (AS29791). Core competency is providing a 100% uptime SLA on your entire stack: ProManaged Hosting Infrastructure, Voxel IP Network, VoxCLOUD Hybrid Cloud and VoxCAST Content Delivery.

Year Founded	1999		Moody's Rating	N/A		Pricing	Y

IaaS Sub-Classification

White Label	
Traditional	
Servers	✓
Storage	
Disaster Recovery	
Backup	
Messaging	
Content Delivery	✓
Desktop-as-a-Service	
Hosting	✓
IT-as-a-Service	
Management Software	

Direct & Channel

None Listed

Summary

Termination Notification (Days)	30
Dedicated Account Management	Y
Live Customer Support	Y
Guaranteed Network Availability	Y
Root/ Administrator Access	Y
Portal Support	Y

Clients

Mochila
Alpari
New York Observer
Pressflex
Twistage

Updates

Available
Q2 2011

CloudSourcing100.com

Regulation and Compliance

N/A

Programming Languages

N/A

Operating Systems Compatibility

Red Hat Enterprise Linux

CLOUD SOURCING 100   Q1 2011

SOLUTION TYPE	PROVIDER TYPE | CUSTOMER CONTROL	
		Wipro www.wipro.com

PROVIDER SUMMARY

Wipro Ltd. is an information technology services corporation headquartered in Bangalore, India. Wipro is one of the largest IT services company in India and employs more than 119,491 people worldwide as of September 2010. Wipro provides clould computing consulting services to enterprises & cloud service originators. Wipro brings a process & vendor agnostic approach to Cloud services through thier system integrating services that cover - Business Process-as-a-Service, Software-as-a-Service, Platform-as-a-Service, and Infrastructure-as-a-Service. In addition, Wipro provides Cloud Security Services and Cloud Testing services span all the four levels of Cloud.

Year Founded	1945	Moody's Rating	N/A	Pricing	N

IaaS Sub-Classification

- White Label
- Traditional
- Servers
- Storage
- Disaster Recovery
- Backup
- Messaging
- Content Delivery
- Desktop-as-a-Service
- Hosting
- IT-as-a-Service
- Management Software

Direct & Channel

None Listed

Clients

None Listed

Summary

Termination Notification (Days)	N/A
Dedicated Account Management	Y
Live Customer Support	Y
Guaranteed Network Availability	N/A
Root/ Administrator Access	N/A
Portal Support	N/A

Updates

Available
Q2 2011

CloudSourcing100.com

Regulation and Compliance

N/A

Programming Languages

N/A

Operating Systems Compatibility

N/A

 **Q1 2011**

SOLUTION TYPE	PROVIDER TYPE	CUSTOMER CONTROL	
			 www.wns.com

PROVIDER SUMMARY

WNS claims that BPaaS is not a new offering, but is a new term for services that companies like WNS have been offering for last 10 years, and was earlier called Platform-based BPO. BPaaS focuses on taking the ASP model (PaaS) to the next level and also provides trained and experienced staff to manage the underlying process, often in a virtual, global and distributed operating model.

Year Founded	1996		Moody's Rating	N/A		Pricing	N

IaaS Sub-Classification	
White Label	
Traditional	
Servers	
Storage	
Disaster Recovery	
Backup	
Messaging	
Content Delivery	
Desktop-as-a-Service	
Hosting	
IT-as-a-Service	
Management Software	

Direct & Channel
None Listed

Summary	
Termination Notification (Days)	N/A
Dedicated Account Management	Y
Live Customer Support	N/A
Guaranteed Network Availability	N/A
Root/ Administrator Access	N/A
Portal Support	N/A

Clients
British Gas
SAS Airlines
Lastminute.com
KLM
T-Mobile

Updates
Available Q2 2011
CloudSourcing100.com

Regulation and Compliance
N/A

Operating Systems Compatibility
N/A

Programming Languages
N/A

ALSBRIDGE

 **Q1 2011**

SOLUTION TYPE

PROVIDER TYPE | CUSTOMER CONTROL

WorkXpress
www.workxpress.com

PROVIDER SUMMARY

WorkXpress is the world's only 5GL PaaS (5th generation software language Platform as a Service). WorkXpress saw early on the value of leading-edge technologies like AJAX, rich JavaScript interfaces & server virtualization, making it one of the earliest providers of real solutions to real businesses based on principles currently recognized as cloud computing. WorkXpress provides more software & better features at a lower cost than traditional software options. The platform offers non-programmers the ability to create limitless sophisticated business applications using five building blocks in an intuitive, drag & drop, point-and-click, secure, web-based environment.

Year Founded	2002		Moody's Rating	N/A		Pricing	Y

IaaS Sub-Classification

White Label	
Traditional	
Servers	
Storage	
Disaster Recovery	
Backup	
Messaging	
Content Delivery	
Desktop-as-a-Service	
Hosting	
IT-as-a-Service	
Management Software	

Direct & Channel

MCF Technology Solution, Collective Intelligence, Visao, LLC, BMG Technology

Clients

EventPro Strategies
R.T. Grim
Servolift
Tex Visions

Summary

Termination Notification (Days)	0
Dedicated Account Management	Y
Live Customer Support	Y
Guaranteed Network Availability	Y
Root/ Administrator Access	N
Portal Support	Y

Updates

Available
Q2 2011

CloudSourcing100.com

Regulation and Compliance

SAS70 TYPE II, PCI/DSS,
Various ISO, HIPPA

Operating Systems Compatibility

Linux Operating Systems

Programming Languages

5GL PaaS

ALSBRIDGE

Proof

Made in the USA
Charleston, SC
27 March 2011